Power Line Worker

Level Three: Transmission

Trainee Guide
Second Edition

PEARSON

Boston Columbus Indianapolis New York San Francisco Upper Saddle River
Amsterdam Cape Town Dubai London Madrid Milan Munich Paris Montreal Toronto
Delhi Mexico City São Paulo Sydney Hong Kong Seoul Singapore Taipei Tokyo

National Center for Construction Education and Research
President: Don Whyte
Director of Product Development: Daniele Stacey
Power Line Worker Project Manager for Transmission: Matt Tischler
Production Manager: Tim Davis
Quality Assurance Coordinator: Debie Ness
Desktop Publishing Coordinator: James McKay
Production Specialist: Megan Casey
Editor: Chris Wilson

Writing and development services provided by Topaz Publications, Liverpool, NY
Lead Writer/Project Manager: Thomas Burke
Desktop Publisher: Joanne Hart
Art Director: Megan Paye
Permissions Editors: Andrea LaBarge, Alison Richmond
Writers: Thomas Burke, Troy Staton, Patricia Vidler

Pearson Education, Inc.
Editorial Director: Vernon R. Anthony
Executive Editor: Alli Gentile
Senior Product Manager: Lori Cowen
Operations Supervisor: Deidra M. Skahill
Art Director: Jayne Conte
Director of Marketing: David Gesell
Executive Marketing Manager: Derril Trakalo
Marketing Manager: Brian Hoehl
Marketing Coordinator: Crystal Gonzalez

Composition: NCCER
Printer/Binder: Document Technology Resources, Fredericksburg, VA
Cover Printer: Document Technology Resources, Fredericksburg, VA
Text Fonts: Palatino and Univers

10 9 8 7 6 5 4 3 2 1

PEARSON

Perfect bound ISBN-13: 978-0-13-294867-8
ISBN-10: 0-13-294867-2

Preface

To the Trainee

Welcome to final year of transmission line worker training. As you enter the workforce, you will discover that many opportunities are available for well-trained line workers. Employment can be found in the fields of natural gas distribution, power generation, and government. Power line workers are expected to be in high demand, as more than 20% of the current workforce will retire by next decade. And the demand will only intensify as the nation struggles to modernize an aging energy infrastructure.

NCCER's *Power Line Worker Level Three: Transmission* covers the skills and knowledge necessary to develop outstanding transmission line workers. Topics addressed include liveline bare-hand construction and maintenance, reconductoring transmission lines, hot stick repair and maintenance, and lift planning. By the end of this training, you will have learned how to properly and safely maintain and repair transmission lines, and how to lift large structures into their final positions.

We wish you success as you progress through this training program. Should you have any comments on how NCCER might improve upon this textbook, please complete the User Update form located at the back of each module and send it to us. We will always consider and respond to input from our customers.

We invite you to visit the NCCER website at **www.nccer.org** for information on the latest product releases and training, as well as online versions of the *Cornerstone* newsletter and Pearson's NCCER product catalog.

Your feedback is welcome. You may email your comments to **curriculum@nccer.org** or send general comments and inquiries to **info@nccer.org**.

NCCER Standardized Curricula

NCCER is a not-for-profit 501(c)(3) education foundation established in 1996 by the world's largest and most progressive construction companies and national construction associations. It was founded to address the severe workforce shortage facing the industry and to develop a standardized training process and curricula. Today, NCCER is supported by hundreds of leading construction and maintenance companies, manufacturers, and national associations. The NCCER Standardized Curricula was developed by NCCER in partnership with Pearson, the world's largest educational publisher.

Some features of the NCCER Standardized Curricula are as follows:

- An industry-proven record of success
- Curricula developed by the industry for the industry
- National standardization providing portability of learned job skills and educational credits
- Compliance with the Office of Apprenticeship requirements for related classroom training *(CFR 29:29)*
- Well-illustrated, up-to-date, and practical information

NCCER also maintains a National Registry that provides transcripts, certificates, and wallet cards to individuals who have successfully completed modules of the NCCER Curricula. *Training programs must be delivered by an NCCER Accredited Training Sponsor in order to receive these credentials.*

Special Features

In an effort to provide a comprehensive, user-friendly training resource, we have incorporated many different features for your use. Whether you are a visual or hands-on learner, this book will provide you with the proper tools to get started in the power line worker industry.

Introduction

This page is found at the beginning of each module and lists the Objectives, Performance Tasks, Trade Terms, and Required Trainee Materials for that module. The Objectives list the skills and knowledge you will need in order to complete the module successfully. The Performance Tasks give you the opportunity to apply your knowledge to the real world duties that power line workers perform. The list of Trade Terms identifies important terms you will need to know by the end of the module. Required Trainee Materials list the materials and supplies needed for the module.

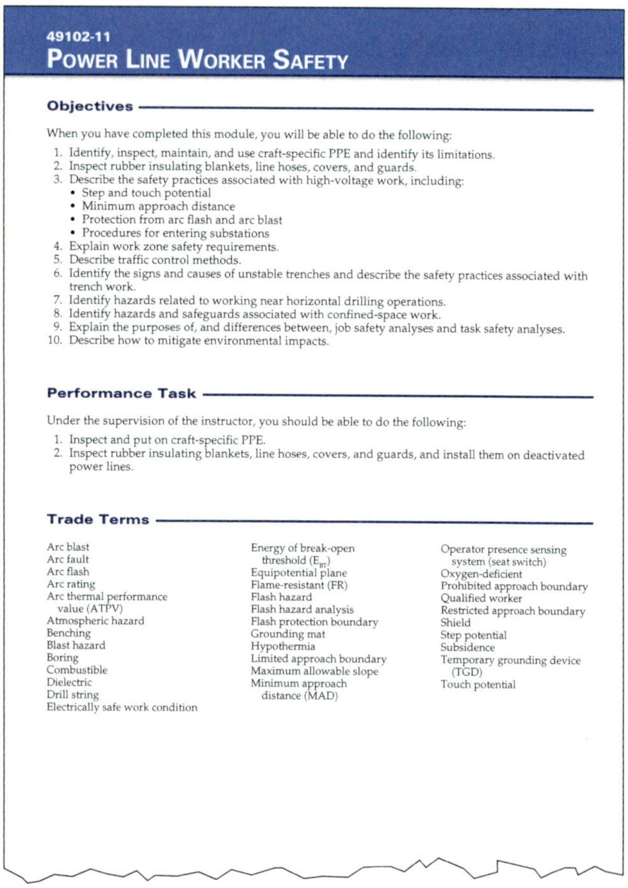

49102-11
POWER LINE WORKER SAFETY

Objectives

When you have completed this module, you will be able to do the following:

1. Identify, inspect, maintain, and use craft-specific PPE and identify its limitations.
2. Inspect rubber insulating blankets, line hoses, covers, and guards.
3. Describe the safety practices associated with high-voltage work, including:
   - Step and touch potential
   - Minimum approach distance
   - Protection from arc flash and arc blast
   - Procedures for entering substations
4. Explain work zone safety requirements.
5. Describe traffic control methods.
6. Identify the signs and causes of unstable trenches and describe the safety practices associated with trench work.
7. Identify hazards related to working near horizontal drilling operations.
8. Identify hazards and safeguards associated with confined-space work.
9. Explain the purposes of, and differences between, job safety analyses and task safety analyses.
10. Describe how to mitigate environmental impacts.

Performance Task

Under the supervision of the instructor, you should be able to do the following:

1. Inspect and put on craft-specific PPE.
2. Inspect rubber insulating blankets, line hoses, covers, and guards, and install them on deactivated power lines.

Trade Terms

Arc blast	Energy of break-open threshold (E_{BT})	Operator presence sensing system (seat switch)
Arc fault	Equipotential plane	Oxygen-deficient
Arc flash	Flame-resistant (FR)	Prohibited approach boundary
Arc rating	Flash hazard	Qualified worker
Arc thermal performance value (ATPV)	Flash hazard analysis	Restricted approach boundary
Atmospheric hazard	Flash protection boundary	Shield
Benching	Grounding mat	Step potential
Blast hazard	Hypothermia	Subsidence
Boring	Limited approach boundary	Temporary grounding device (TGD)
Combustible	Maximum allowable slope	Touch potential
Dielectric	Minimum approach distance (MAD)	
Drill string		
Electrically safe work condition		

Color Illustrations and Photographs

Full-color illustrations and photographs are used throughout each module to provide vivid detail. These figures highlight important concepts from the text and provide clarity for complex instructions. Each figure reference is denoted in the text in *italic type* for easy reference.

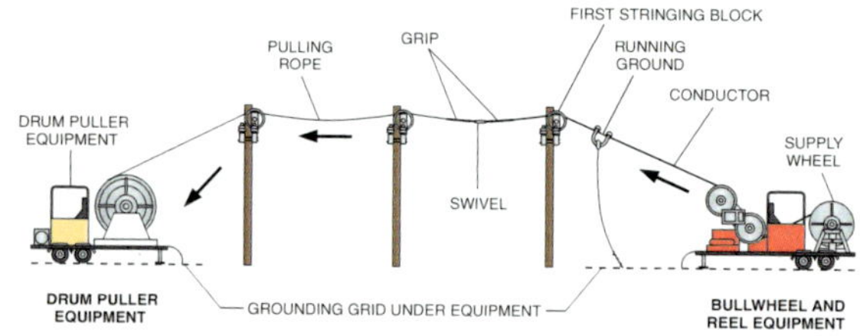

Figure 38 Pulling conductors.

Notes, Cautions, and Warnings

Safety features are set off from the main text in highlighted boxes and are organized into three categories based on the potential danger of the issue being addressed. Notes simply provide additional information on the topic area. Cautions alert you of a danger that does not present potential injury but may cause damage to equipment. Warnings stress a potentially dangerous situation that may cause injury to you or a co-worker.

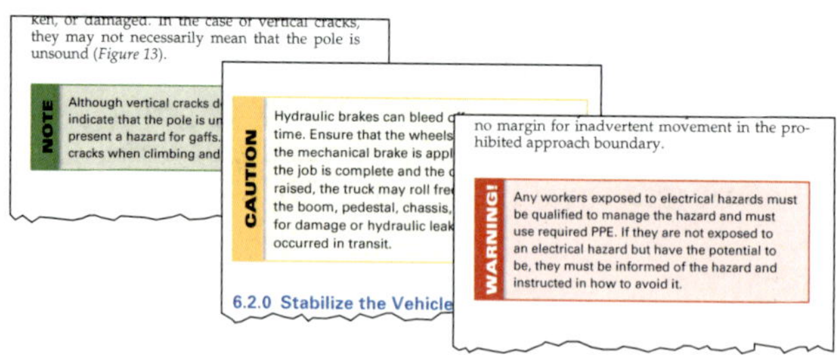

Did You Know?

The Did You Know? features offer hints, tips, and other helpful bits of information from the trade.

Did You Know?

Approximately 30,000 nonfatal and 1,000 fatal electrical shock accidents occur each year. However, the majority of hospital admissions due to electrical accidents are from arc flash burns, not from shocks. An arc flash releases thermal energy at temperatures up to 35,000°F.

Source: *NFPA 70E®, Electrical Safety in the Workplace, Annex K.*

On Site

On Site features provide a head start for those entering the electrical transmission and distribution fields by presenting technical tips and professional practices from power line workers on a variety of topics. On Site features often include real-life scenarios similar to those you might encounter on the job site.

On Site

Transformers

Large distribution transformers at power substations step down the power to the level required for local distribution. Pole transformers, such as the one shown here, step the power down further to the voltages needed for homes and businesses.

49103-11_SA02.EPS

Think About It

Think About It features use "What if?" questions to help you apply theory to real-world experiences and put your ideas into action.

Think About It

The Magic of Electricity

The effect of the flow of electrons occurs at a speed that is close to the speed of light, about 186,000 miles per second. How long does it take the light from the end of a flashlight to reach the floor? If you ran a light circuit from Maine to California and flipped the switch, how long would it take for the light to come on?

Step-by-Step Instructions

Step-by-step instructions are used throughout to guide you through technical procedures and tasks from start to finish. These steps show you not only how to perform a task but also how to do it safely and efficiently.

trically safe work condition. Most companies have detailed written procedures for performing this task.

Step 1 Determine whether any other crews are working on the circuit. All distribution lines are treated as if they are energized unless your team has performed a de-energizing procedure. If two or more crews are working on the same lines or equipment, each crew must independently perform a de-energizing procedure and apply their own lockout/tagout devices to energy controls.

Step 2 Designate one qualified member of the crew as the employee in charge of the electrical clearance.

Trade Terms

Each module presents a list of Trade Terms that are discussed within the text and defined in the Glossary at the end of the module. These terms are denoted in the text with **bold, blue type** upon their first occurrence. To make searches for key information easier, a comprehensive Glossary of Trade Terms from all modules is located at the back of this book.

You will hear the term *circuit* throughout your training. An electrical circuit contains, at minimum, a **voltage** source, a load, and **conductors** (wires) to carry the electrical **current** (*Figure 1*). The circuit should also have a means to stop and start the current, such as a switch.

Electricity is concerned with cause and effect. The presence of voltage (volts) in a closed circuit causes current (amps) to flow. The more voltage you apply, the more current will flow. However, the amount of current flow is also determined by how much **resistance**, in **ohms** (Ω), the load offers

Review Questions

Review Questions are provided to reinforce the knowledge you have gained. This makes them a useful tool for measuring what you have learned.

Review Questions

1. All of the following items are considered by OSHA to be acceptable fall prevention systems for working at heights above 6 feet *except* for _____.
   a. a guardrail system
   b. personal fall protection
   c. the two-climber buddy system
   d. a safety net system

2. Fall arrest systems _____.
   a. stop or take control of a fall in progress
   b. prevent falls from ever occurring
   c. ensure that the climber does not swing from side-to-side during a fall
   d. ensure that every fall is reported to the proper authorities

3. OSHA requires that a PFAS limits the maximum force imparted to the body during fall arrest to _____.
   a. 500 pounds
   b. 1,200 pounds
   c. 1,800 pounds
   d. 2,500 po

7. When an anchor point rated at 5,000 pounds is not available, the climber must then ensure that the chosen anchor position can at least withstand the maximum load that could be placed on it during a fall arrest, multiplied by a factor of _____.
   a. 1.5
   b. 2
   c. 3
   d. 5

8. Anchor points that are used as fall restraint anchors, and *not* as fall arrest anchors, must be load rated to handle _____.
   a. 2,500 pounds
   b. 3,000 pounds
   c. 5,000 pounds
   d. 7,500 pounds

9. The climber's personal fall arrest harness must be connected to the anchor point using _____.
   a. at least two lanyards

NCCER Curricula

NCCER's training programs comprise more than 80 construction, maintenance, pipeline, and utility areas and include skills assessments, safety training, and management education.

Boilermaking
Cabinetmaking
Carpentry
Concrete Finishing
Construction Craft Laborer
Construction Technology
Core Curriculum:
  Introductory Craft Skills
Drywall
Electrical
Electronic Systems Technician
Heating, Ventilating, and
  Air Conditioning
Heavy Equipment Operations
Highway/Heavy Construction
Hydroblasting
Industrial Coating and Lining
  Application Specialist
Industrial Maintenance
  Electrical and Instrumentation
  Technician
Industrial Maintenance
  Mechanic
Instrumentation
Insulating
Ironworking
Masonry
Millwright
Mobile Crane Operations
Painting
Painting, Industrial
Pipefitting
Pipelayer
Plumbing
Reinforcing Ironwork
Rigging
Scaffolding
Sheet Metal
Signal Person
Site Layout
Sprinkler Fitting
Tower Crane Operator
Welding

Green/Sustainable Construction

Building Auditor
Fundamentals of Weatherization
Introduction to Weatherization
Sustainable Construction
  Supervisor
Weatherization Crew Chief
Weatherization Technician
Your Role in the Green
  Environment

Energy

Alternative Energy
Introduction to the Power
  Industry
Introduction to Solar
  Photovoltaics
Introduction to Wind Energy
Power Industry Fundamentals
Power Generation Maintenance
  Electrician
Power Generation I&C
  Maintenance Technician
Power Generation Maintenance
  Mechanic
Power Line Worker
Power Line Worker: Distribution
Power Line Worker: Substation
Power Line Worker:
  Transmission
Solar Photovoltaic Systems
  Installer
Wind Turbine Maintenance
  Technician

Pipeline

Control Center Operations,
  Liquid
Corrosion Control
Electrical and Instrumentation
Field Operations, Liquid
Field Operations, Gas
Maintenance
Mechanical

Safety

Field Safety
Safety Orientation
Safety Technology

Management

Fundamentals of Crew
  Leadership
Project Management
Project Supervision

Supplemental Titles

Applied Construction Math
Careers in Construction
Tools for Success

Spanish Translations

Basic Rigging
  (Principios Básicos de
  Maniobras)
Carpentry Fundamentals
  (Introducción a la
  Carpintería, Nivel Uno)
Carpentry Forms
  (Formas para Carpintería,
  Nivel Trés)
Concrete Finishing, Level One
  (Acabado de Concreto,
  Nivel Uno)
Core Curriculum:
  Introductory Craft Skills
  (Currículo Básico:
  Habilidades Introductorias del
  Oficio)
Drywall, Level One
  (Paneles de Yeso, Nivel Uno)
Electrical, Level One
  (Electricidad, Nivel Uno)
Field Safety
  (Seguridad de Campo)
Insulating, Level One
  (Aislamiento, Nivel Uno)
Ironworking, Level One
  (Herrería, Nivel Uno)
Masonry, Level One
  (Albañilería, Nivel Uno)
Pipefitting, Level One
  (Instalación de Tubería
  Industrial, Nivel Uno)
Reinforcing Ironwork, Level One
  (Herreria de Refuerzo,
  Nivel Uno)
Safety Orientation
  (Orientación de Seguridad)
Scaffolding
  (Andamios)
Sprinkler Fitting, Level One
  (Instalación de Rociadores,
  Nivel Uno)

Acknowledgments

This curriculum was revised as a result of the farsightedness and leadership of the following sponsors:

Cianbro Corporation
MasTec, Inc.
Pike Electric, LLC

This curriculum would not exist were it not for the dedication and unselfish energy of those volunteers who served on the Authoring Team. A sincere thanks is extended to the following:

Don Daniel
Terry Malloy
David McPeak

NCCER Partners

American Fire Sprinkler Association
Associated Builders and Contractors, Inc.
Associated General Contractors of America
Association for Career and Technical Education
Association for Skilled and Technical Sciences
Carolinas AGC, Inc.
Carolinas Electrical Contractors Association
Center for the Improvement of Construction Management and Processes
Construction Industry Institute
Construction Users Roundtable
Construction Workforce Development Center
Design Build Institute of America
Merit Contractors Association of Canada
Metal Building Manufacturers Association
NACE International
National Association of Minority Contractors
National Association of Women in Construction
National Insulation Association
National Ready Mixed Concrete Association
National Technical Honor Society
National Utility Contractors Association
NAWIC Education Foundation
North American Technician Excellence
Painting & Decorating Contractors of America
Portland Cement Association
Skills USA
Steel Erectors Association of America
The Manufacturers Institute
U.S. Army Corps of Engineers
University of Florida, M. E. Rinker School of Building Construction
Women Construction Owners & Executives, USA

NCCER Business Partners

Contents

Module Four

Lift Planning

Explains how the load weight and center of gravity affect a lift. Covers sling selection and spreader bar use, as well as the use of cribbing and inclined planes. Includes case studies from three complex lifts. (Module ID 38302-11; 40 Hours)

Glossary

Index

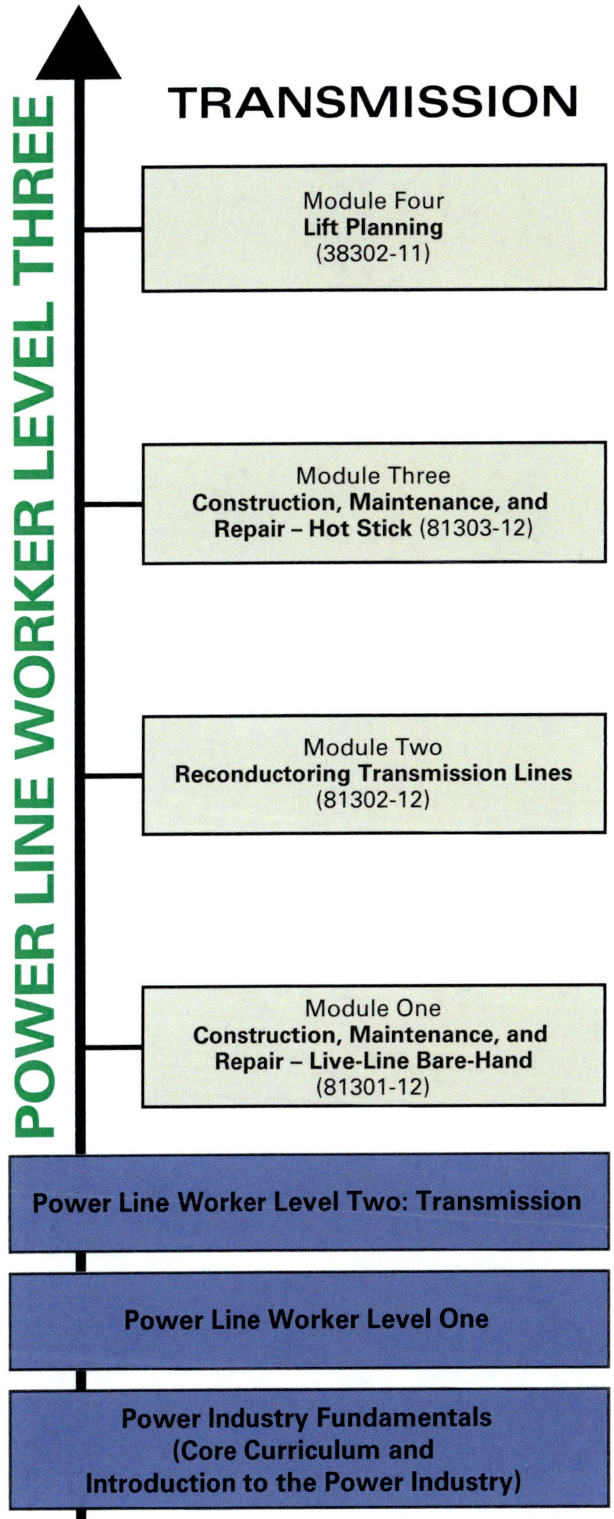

This course map shows all of the modules in *Power Line Worker Level Three: Transmission*. The suggested training order begins at the bottom and proceeds up. Skill levels increase as you advance on the course map. The local Training Program Sponsor may adjust the training order.

81301-12

Construction Maintenance and Repair – Live-Line Bare-Hand

Module One

Trainees with successful module completions may be eligible for credentialing through NCCER's National Registry. To learn more, go to www.nccer.org or contact us at **1.888.622.3720**. Our website has information on the latest product releases and training, as well as online versions of our *Cornerstone* newsletter and Pearson's product catalog.

Your feedback is welcome. You may email your comments to **curriculum@nccer.org,** send general comments and inquiries to **info@nccer.org**, or fill in the User Update form at the back of this module.

81301-12
CONSTRUCTION MAINTENANCE AND REPAIR – LIVE-LINE BARE-HAND

Objectives

When you have completed this module, you will be able to do the following:

1. Identify safety practices associated with live-line bare-hand work.
2. Identify the required personal protective equipment (PPE) used specifically in live-line bare-hand work.
3. Describe basic procedures for initiating live-line bare-hand tasks, such as:
   - Bonding to an energized line
   - Care and preparation of a bucket truck boom
   - Donning the required PPE

Performance Tasks

Under the supervision of your instructor, you should be able to do the following:

1. Don and electrically test a conductive suit, including gloves and footwear.
2. Demonstrate how a worker bonds a conductive suit and a bucket to an energized line.
3. Demonstrate how to complete a current leakage test of a bucket boom.

Trade Terms

Ergonomic component
Hold off

Industry Recognized Credentials

If you're training through an NCCER-accredited sponsor you may be eligible for credentials from NCCER's Registry. The ID number for this module is 81301-12. Note that this module may have been used in other NCCER curricula and may apply to other level completions. Contact NCCER's Registry at 888.622.3720 or go to **nccer.org** for more information.

Contents

Topics to be presented in this module include:

Figures and Tables

1.0.0 Introduction

Transmission lines carry electrical power from the point of generation to the many subtransmission lines and substations scattered across the country. From these points, power is distributed to consumers of every size. Since transmission lines form the backbone of the national grid, a failure of any kind represents a problem to many people and facilities. This fact has driven the development of techniques to repair and maintain transmission lines while they remain energized.

Power distribution systems operate at significantly lower voltages than transmission systems. The predominant method of working on hot distribution systems is referred to as rubber-gloving. Hot sticks are also used, allowing workers to stay at a safe distance from energized components and conductors. Many products, including the rubber gloves, have been developed that fully protect workers from distribution system voltages. With rubber gloves that are rated for the voltage present, workers can handle energized components safely. However, the safety equipment used in rubber gloving cannot protect against the extreme voltages present in many transmission systems. Touching a 765kV conductor with the highest-rated rubber glove, while also in contact with a grounded object, still results in disaster. Although hot sticks are used safely for transmission line work, there are a number of tasks that simply cannot be completed using a long pole.

Since known and available PPE cannot provide sufficient protection from high voltages, another approach is needed. Rather than avoiding contact with an energized transmission line, live-line bare-hand workers use the opposite tactic. In bare-hand work, workers freely contact and embrace the energized surface (*Figure 1*), while ensuring that no contact is made with a grounded object or another phase of power. The focus then, is on isolating the worker from these surfaces rather than the energized surfaces. The bare-hand worker becomes a part of the electrical circuit.

This module presents the basic safety procedures required to service energized transmission systems. The required PPE is covered, as well as the basic steps to access the work location and complete the initial electrical bond. However, trainees should keep in mind that transmission system components and work procedures differ among employers and utilities. Always follow the specific procedures provided by your employer for the area of work and the type of service being performed.

2.0.0 Live-Line Safety

Live-line work on transmission systems can certainly be hazardous. Few people would debate that fact. However, many man-hours are logged each year by live-line workers without a serious incident. As is the case with many trades, good training, safe working habits, and attention to de-

81301-12_F01.EPS

Figure 1 Bare-hand insulator inspection.

On Site

Bare-Hand History

The original bare-hand procedures were tested and developed by Harold Rorden, an engineer working for American Electric Power, in 1960. Assisted by another engineer employed by Ohio Brass, Whitey White, Rorden began testing at 4kV and eventually went as high as 380kV. "It seems a reckless folly," Rorden himself commented to *Popular Mechanics* in 1961. "But it is not essentially different from being at ground potential and insulated from all other energized conductors." Field tests included the replacement of an insulator and armor rods on the 34kV distribution line that fed White's home. With this boost in confidence in their techniques, the duo repaired two 138kV transmission conductors damaged by rifle bullets. (Even in 1961, significant threats to the integrity of transmission lines included bored hunters and firearm-equipped vandals.)

tail are the important factors in successful bare-hand work.

One rule is often repeated in bare-hand work: If you are not electrically connected to it, don't touch it. A bond must be formed with any conductive object before it is handled. Non-conductive ropes and hot sticks must be used to move tools or parts between workers bonded to the conductor and those who are at ground potential. A bonding process must take place between the energized worker and the tool or part. Line workers must think carefully before reaching out for anything and consider its ground potential. Bonding is covered later in this module in greater depth.

Although other parts of *OSHA Standard 1910.269* apply to live-line bare-hand work, *Standard 1910.269(q)(3)* pertains specifically to bare-hand work. This portion outlines the training required and provides directives for the safe execution of bare-hand work.

One provision is that bare-hand work may not be performed when adverse weather is present. Examples of adverse weather provided are thunderstorms, high winds, snowstorms, and ice storms. Wind is a specific concern, as it causes the movement of conductors. As a conductor moves with the wind, it may intermittently enter the MAD of a line worker who would otherwise be a safe distance away. The job site supervisor is the individual who typically must determine if weather conditions are sufficiently adverse to stop or delay work.

2.1.0 Job Safety Analysis

Due to the diversity of tasks and the different configurations of structures and components, a Job Safety Analysis (JSA) is always recommended. This review of the procedures, also known as a Job Hazard Analysis, is required by most utilities and employers. Once the job begins, issues may arise that require procedural changes. They can usually be addressed and remedied through a discussion between experienced workers and the on-site supervisor. During that time, a work stoppage is generally required to ensure all parties understand the change. Changes that reflect a safer and/or more efficient process may become part of the JSA and future procedural guides.

The JSA is usually done in advance of the work. It is then reviewed and discussed along with the procedures to be followed for the day as part of the pre-job briefing. Some refer to this pre-job meeting as a tailboard, or tailgate, meeting since it is often held somewhere near the work vehicles.

2.2.0 Minimum Approach Distances

Table 1 shows the minimum approach distances (MADs) for both phase-to-phase and phase-to-ground energized surfaces. The phase-to-ground MAD is significantly less since the voltage potential is lower from one phase to ground than it is between two phases of power. In wye systems, the phase-to-ground voltage can be determined by dividing the phase-to-phase voltage by 1.732. Although the MADs assigned are affected by other factors, the distances are roughly proportional to the difference in voltage.

The MAD refers to the distance that qualified personnel must maintain from an energized surface without proper PPE. However, the concept for bare-hand work is somewhat different. For tasks other than bare-hand, the challenge is to maintain the MAD from all energized surfaces; contact with anything at ground potential is expected and necessary. In bare-hand work though, the worker becomes an energized surface. Here, the challenge is to maintain the MAD from ground or another phase of power. The essential rule for working bare-hand does remain the same—avoid any two points of contact where current can both enter and exit the body. Of course, at typical transmission voltages, actual contact is not necessary for disaster to occur. Passing through the invisible MAD barrier should essentially be considered contact by a line worker. The MAD must be maintained at all times and under all circumstances.

Table 1 Minimum Approach Distances

Kilovolts (AC)	Phase-to-Ground	Phase-to-Phase
0.05 to 1.0	Avoid contact	Avoid contact
1.1 to 15.0	2'-1" (0.64m)	2'-2" (0.66m)
15.1 to 36.0	2'-4" (0.72m)	2'-7" (0.77m)
36.1 to 46.0	2'-7" (0.77m)	2'-10" (0.85m)
46.1 to 72.5	3'-0" (0.90m)	3'-6" (1.05m)
72.6 to 121	3'-2" (0.95m)	4'-3" (1.29m)
138 to 145	3'-7" (1.09m)	4'-11" (1.50m)
161 to 169	4'-0" (1.22m)	5'-8" (1.71m)
230 to 242	5'-3" (1.59m)	7'-6" (2.27m)
345 to 362	8'-6" (2.59m)	12'-6" (3.80m)
500 to 550	11'-3" (3.42m)	18'-1" (5.50m)
765 to 800	14'-11" (4.53m)	26'-0" (7.91m

From *OSHA 29 CFR 1910.269 Electric Power Generation, Transmission, and Distribution*

2.2.1 Working on Outer Phases

When a worker is servicing live lines near a pole or tower, the most common approach is with a bucket truck. In *Figure 2*, note the various surfaces that must remain outside of the MAD.

When working near multiple points of danger, such as two grounded surfaces, the worker's area should be considered an 8-foot sphere. The center of the sphere is near the workers waistline. The MAD to each surface is measured to the perimeter of the imaginary sphere. Since the lower portion of the bucket truck boom is typically uninsulated, the phase-to-ground MAD must be maintained from the worker sphere to the lower boom. In addition, the worker's sphere must remain outside of the phase-to-ground MAD from any portion of the structure itself. When working on outer phases, maintaining the phase-to-phase MAD is generally not difficult.

2.2.2 Working on Inner Phases

When working on inner phases, maintaining the proper MAD becomes more complex. Refer to *Figure 3*. The worker's sphere remains at 8 feet in diameter. The phase-to-ground MAD must be maintained between the sphere and the uninsulated lower boom. The phase-to-ground MAD must also be maintained between the worker's sphere and any portion of the supporting structure.

When the worker is positioned somewhere between an energized surface and a grounded surface, then the total distance in each direction must be equal to or greater than the phase-to-ground MAD. Again, the MAD to each surface is measured to the perimeter of the imaginary sphere. These measurements are shown in *Figure 3* as MAD_{D1} and MAD_{D2}. For example, if the phase-to-ground MAD is 8.5 feet, then the sum of MAD_{D1} and MAD_{D2} must be equal to or greater than 8.5 feet. If MAD_{D1} is 3 feet, MAD_{D2} must be 5.5 feet or more. These same concepts can be applied to the phase-to-phase MAD.

2.2.3 Altitude Correction

The MADs must be adjusted as altitude increases above 3,000 feet per *OSHA Standard 1910.269, Table R-10*. See *Table 2* for multipliers that are to be applied at higher altitudes.

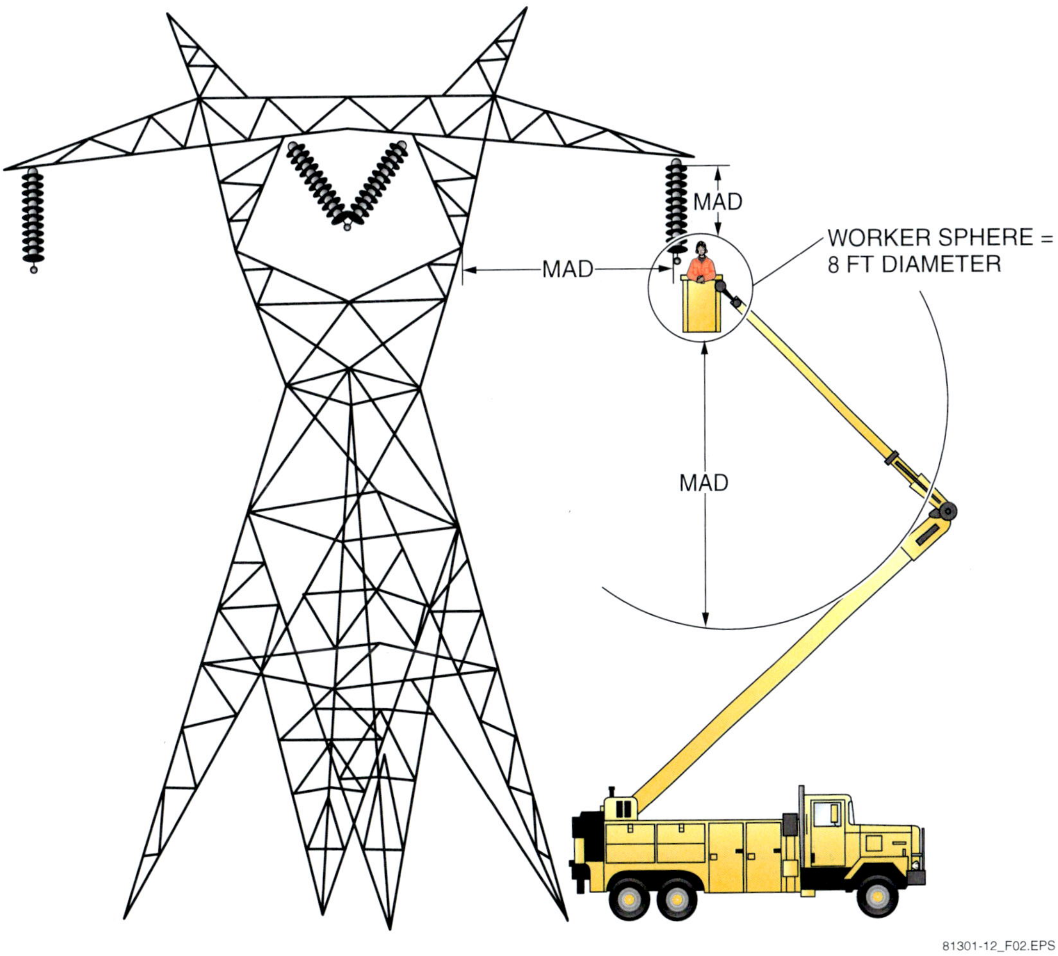

Figure 2 MAD scenario—working an outer phase.

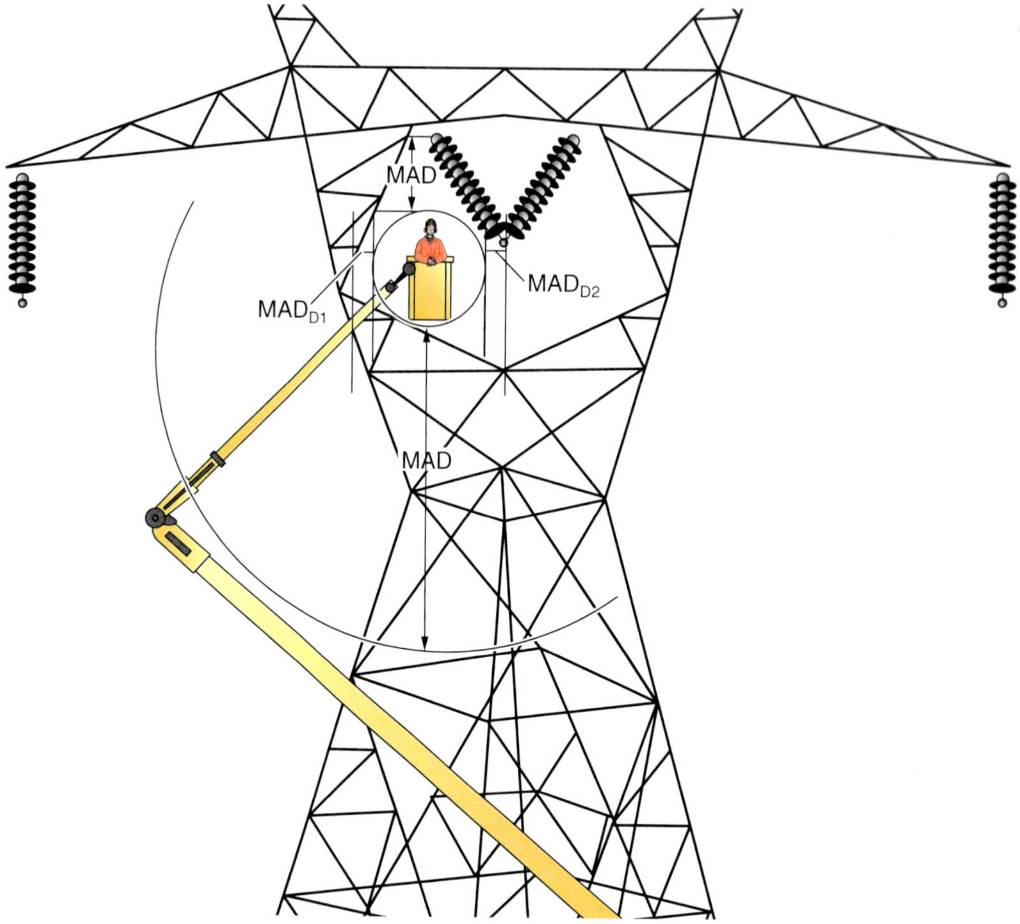

Figure 3 MAD scenario—working an inner phase.

Table 2 MAD Altitude Correction Multipliers

Altitude	Correction Factor
3,000' (900m)	1.00
4,000' (1,200m)	1.02
5,000' (1,500m)	1.05
6,000' (1,800m)	1.08
7,000' (2,100m)	1.11
8,000' (2,400m)	1.14
9,000' (2,700m)	1.17
10,000' (3,000m)	1.20
12,000' (3,600m)	1.25
14,000' (4,200m)	1.30
16,000' (4,800m)	1.35
18,000' (5,400m)	1.39
20,000' (6,000m)	1.44

From *OSHA Standard 1910.269(l)(10), Table R-10*

One factor in the calculation of MADs is a value that allows for inadvertent or unintentional movement. OSHA refers to this added distance as the **ergonomic component** of the MAD. This is a value that is added to the electrical component of each MAD in *OSHA Standard 1910.269(l)(10), Table R-6.* The ergonomic component is added to compensate for inadvertent movement toward an energized surface, and the time it may take a human to stop his or her motion in the wrong direction. For all voltages above 72.5kV, the distance added to the electrical component is 1 foot. At 72.5kV and lower, the distance added is 2 feet. One little known fact is that the ergonomic component of the MAD for voltages less than 72.5kV is greater than the electrical component. These added ergonomic components are already reflected in OSHA's MAD table.

Altitude correction multipliers are not to be applied directly to the values in the MAD table. The ergonomic component must first be removed, and then the multiplier is applied to the electrical component. After multiplying the electrical component, the ergonomic component is added back in to arrive at the total MAD.

The following is an example of how altitude correction is applied. Assume that live-line work is to be done on towers through a mountain pass at an altitude of 6,000 feet, and the phase-to-phase voltage is 345kV. The MAD for this voltage is 8.5 feet phase-to-ground and 12.5 feet phase-to-phase.

First, remove the ergonomic component of 1 foot from the two values:

8.5 feet – 1 foot = 7.5 feet electrical component of 345kV MAD, phase-to-ground

12.5 feet – 1 foot = 11.5 feet electrical component of 345kV MAD, phase-to-phase

Next, locate the altitude correction factor in *Table 2* for 6,000 feet. The multiplier from the table is 1.08. Apply the multiplier to the calculated electrical components:

$$7.5 \text{ feet} \times 1.08 = 8.1 \text{ feet}$$

$$11.5 \text{ feet} \times 1.08 = 12.42 \text{ feet}$$

Finally, to determine the actual MAD for 345kV at 6,000 feet, add the ergonomic component of 1 foot back to the adjusted electrical components:

8.1 feet + 1 foot = 9.1 feet MAD phase-to-ground for 345kV @ 6,000 feet

12.42 feet + 1 foot = 13.42 feet MAD phase-to-phase for 345kV @ 6,000 feet

2.3.0 Live-Line PPE

PPE associated with bare-hand live-line work is significantly different from that required for other approaches. However, clothing that contains acetate, nylon, polyester, or rayon is prohibited because of its potential to burn, melt, or smolder. Clothing made from these materials is never a good choice for electrical work.

2.3.1 Conductive Suits

Bare-hand live-line work can be accomplished at any voltage. At distribution system voltages however, it is rarely used. A rubber-glove approach is typically used instead. Between 69kV and 115kV, the use of a conductive suit is optional. At 115kV and up, the current flow through the human body begins to become noticeable and somewhat debilitating. As a result, conductive suits are generally used at or above 115kV.

Conductive suits (*Figure 4*) are typically constructed of Nomex™ or similar aramid-based fibers for fire resistance, woven together with minute stainless steel fibers. Conductive suits generally have serial numbers to track their history.

The suit provides a conduit for the electrical energy, while the worker remains safe inside from the electrical field. Although the amount of current flow through the body would usually not be sufficient to cause injury without the suit, it can certainly interfere with movement and create an uneasy feeling for the worker.

Conductive suits are based on the Faraday cage principle. Based on Michael Faraday's work in the early 1800s, electrical energy applied to a conductive cage negates any effect to the contents of the cage. Aircraft create the same effect when struck by lightning in flight. Faraday's principle is proven here as well, as the occupants of the aircraft are fully protected from the electrical energy.

Conductive suits can be one-piece coverall style, or two-piece. When two pieces are used, conductive straps are provided inside the pants and jacket so that they remain electrically teth-

81301-12_F04.EPS

Figure 4 Conductive suit.

ered. They must be securely tied together at all times. Bonding tethers are also attached to the garment so that it can be connected to the bucket or energized surface. A hood should also be part of the suit, generously sized to accommodate a hard hat.

Unused tethers should always be rolled up, taped, and stowed inside protective pockets. When two or more workers are together, they should either bond their suits together, or bond with the energized surface at the same location if possible. This ensures that they are both at the same electrical potential.

It should be noted that a worker's body and the suit become energized before contact is made with the energized surface, due to induction. Once the lineman is close, the difference in voltage potential may only be 10kV to 50kV just before contact is made. This is still a sufficient difference in potential to cause an arc to develop as contact is initiated. If not for the induction process that brings the worker closer to the transmission voltage, the arc would be far more pronounced.

The resistance of the suit must remain very low to ensure that current flows through it easily, leaving the worker inside at a higher resistance to current flow. The suit, as well as conductive gloves and socks, can be tested with an ohmmeter. The basic concept for testing is to test various points of the garments that are farthest apart. After donning the suit, gloves, and footwear, a continuity test is typically done between the following:

- The end of the jacket bonding tether and the shoe sole
- One sleeve cuff and the other cuff
- One pant leg cuff and the other
- One pant leg cuff and the opposite sleeve cuff
- The hood and the shoe sole
- One glove and the other

The ohmmeter should never read a resistance higher than 100,000 ohms. A lower pass/fail threshold will likely be established by the utility or other employer. If the reading is higher, a worker should first check to ensure the garments are properly bonded to each other. If that does not cure the problem, then begin checking individual garments or sections to determine which piece or section may be faulty.

Conductive garments and gear must be stored in a dry environment. Follow the manufacturer's instructions for laundering. Some suits and gear may require dry cleaning rather than laundering. For most products, washing in a standard washing machine is recommended, using lukewarm or cold water. They should be washed on a gentle or delicate cycle to avoid stressing sections of the suit. Any mild detergent is typically acceptable. Although drying conductive garments in a clothes dryer is acceptable, it is recommended that they only be allowed to remain in the dryer about 10 minutes. They are then removed and placed on a drying rack or corrosion-resistant hanger to complete the drying process.

Damaged garments or their components should be replaced before use. Intermittent discharges or arcs felt by the worker are a likely indication that the suit or other garment is defective.

2.3.2 Gloves

Conductive gloves are an important part of the PPE. At least one lineman has experienced the effects of placing a completely bare hand on a 230kV conductor. Although uninjured, the worker described the effect as a noticeable numbness that persisted for roughly ten minutes. The conductive gloves allow the initial balancing of potentials to occur around the workers hand.

To further protect the worker from arc-induced discomfort, a wand is typically used to make the initial contact with an energized surface.

Gloves should be hand-washed in lukewarm or cold water. Once they are rinsed and all water squeezed out, they are allowed to air-dry.

2.3.3 Footwear

Conductive footwear (*Figure 5*) is required for bare-hand work as well. Conductive socks should also be worn to remove any electrical barriers between the body and the shoe. Even foot powders

81301-12_F05.EPS

Figure 5 Conductive work shoe.

should not be used with conductive footwear, as some electrical resistance could be added. Most conductive footwear includes a strap bonded to the shoe that wraps around the leg to ensure good conductivity. They are usually identified with some sort of visible exterior label, indicating that the shoes are conductive. When conductive socks are used, the shoe strap is not generally required.

All footwear designed to prevent workplace injury, including conductive footwear, should meet *ASTM International Standard F2413-11, Standard Specification for Performance Requirements for Protective (Safety) Toe Cap Footwear.* Conductive footwear meeting this standard is marked CD. Like gloves, conductive socks should be hand-washed and allowed to air-dry.

2.3.4 Hard Hats

For bare-hand live-line work, hard hats do not need to be as conductive as the suit itself. Although Class E hard hats are typically worn to provide electrical insulation, bare-hand workers do not desire this protection. As a result, Class C hard hats can be worn while bare-hand work is in progress. Class C hard hats are not tested electrically to determine their conductivity, but are classified as conductive regardless. The hood of the conductive suit should cover the hard hat, and be kept from falling into the worker's eyes by the brim of the hard hat.

2.4.0 Reclosers

A transmission or subtransmission system may be equipped with one or more reclosers (*Figure 6*). Reclosers operate automatically by opening when a fault is detected, then reclosing the circuit after a predetermined period of time. This may occur several times if the condition causing the initial fault has not been cleared. Reclosers help minimize common outages that result from minor and short-lived fault conditions, such as a fallen tree limb.

During live-line work, the automatic function of all reclosers must be disabled. If a fault occurs during the work, especially one involving a line worker, it is certain that re-energizing the circuit repeatedly is not desirable. The request for, and subsequent confirmation of, disabled reclosers is typically referred to as a **hold off**. Other utilities and employers may refer to it as a one-shot request. The protective relays that control reclosers can sometimes be programmed remotely for one-shot operation, while others must be done manually at the installation location.

81301-12_F06.EPS

Figure 6 800kV circuit breaker/recloser.

3.0.0 Live-Line Equipment

There are few special tools that are specific to bare-hand work. The equipment used is virtually identical to that used for hot stick work. This includes the various types of hot sticks, ladders, and strain poles and carriers. The bucket truck is also common to both hot stick and bare-hand work. Beyond these items, most of the tools used by bare-hand workers are common hand tools.

Keep in mind that bare-hand work does not necessarily mean that all workers are working bare-hand. For example, a large suspension insulator may be worked by a bare-hand worker at the hot end, while a crew works from above with hot sticks. The workers remain isolated from each other by the length of the insulator and the use of fiberglass-reinforced plastic (FRP) sticks and poles.

3.1.0 Ladders and Work Platforms

Insulated ladders are an important component of live-line work. They can be used in many different ways beyond the vertical position to access the work site.

For all ladders used in the vertical position, safety ropes must be provided on both sides of the ladder. They should extend from the top to at least the third step from the bottom. The safety ropes provide a point of connection for the worker's fall arrest lanyard.

Hook ladders (*Figure 7*) are used in a variety of ways. The ladder can be hung from a conductor, pole, or tower structural member. The chains are used to positively secure the ladder. Eyes near the opposite end can be used to secure the bottom to reduce movement and swinging. Three-rail mod-

81301-12_F07.EPS

Figure 7 Hook ladder.

els (*Figure 8*) are stronger and more rigid for use in a horizontal plane. Where a longer ladder is needed that would normally be too long to transport and handle, a spliced ladder can be used (*Figure 9*). The rails are spliced together using steel couplings to assemble two or more sections.

With proper support and planning, insulated ladders can be used in many different ways. *Figure 10* shows a worker using a ladder to access the wire tongs holding the bottom of a strain carrier assembly away from the tower. Another ladder is being used horizontally.

Insulated ladders should be tested daily for current leakage, and any time they are moved for use at a higher voltage. They should be tested for one minute in contact with an energized surface. The current measured must be less than 0.33 microamperes (μA) per kV of phase-to-ground voltage. Test results should be recorded and kept with the ladder. Ladders must be taken out of service for more extensive testing and inspection every two years.

81301-12_F08.EPS

Figure 8 Three-rail ladder.

81301-12_F09.EPS

Figure 9 Spliced ladder.

81301-12_F10.EPS

Figure 10 Insulated ladders in use.

Insulated ladder rungs have sand mixed into the finish for improved grip. If the rung surface becomes worn to the point where grip is lost, the ladder should be removed from service.

Work platforms (*Figure 11*) provide a secure place for line workers to reduce climbing stress and work more comfortably and safely. They can be attached to common poles or to steel towers with the appropriate hardware.

3.2.0 Conductor Pulling Tools

Pulling grips (*Figure 12*) are used to grasp conductors without slippage, so they can be pulled to the proper tension. Pulling grips are also used when conductors must be cut and spliced, or cut for attachment to a dead end.

A ratcheting chain hoist (*Figure 13*) is often used to apply the desired tension on the conductor while repairs are made. One end is attached to the pulling grip, and the other to a suitable point on the structure. To isolate the energized conductor from the hoist and provide added length, a strain link stick (*Figure 14*) can be connected to the pulling grip. They are available in various lengths. Strain link sticks provide the necessary distance and electrical protection for a number of conductor-rigging chores. They are also available with an integral clamping device or rollers to allow the conductor some freedom of movement.

3.3.0 Live-Line Rope

Rope is necessary for a number of rigging and conductor-control tasks. Rope used in live-line work (*Figure 15*) needs to have a high dielectric strength and resistance to water absorption. For this reason, natural fiber ropes are not acceptable.

81301-12_F11.EPS

Figure 11 Work platform.

81301-12_F12.EPS

Figure 12 Pulling grip.

Although rope is generally a poor conductor, water absorption can turn it into a serious hazard.

Make sure that all rope used during live-line work is designed for the purpose, and is kept clean and dry at all times. Live-line rope should be identified as such, and should undergo scheduled retesting. Rope should be tested before use each day. This is often done by using a hot stick to push the rope into contact with an energized conductor. A portable unit capable of testing hot sticks in the field can also be used to test rope.

Store the rope in a good container with desiccant materials to ensure it remains dry. All other accessories should also be kept clean and dry. It is best to use tarps or other material in the work area to keep the rope from lying on the ground where it can absorb moisture.

3.4.0 The Bucket Truck

The bucket truck is an integral part of live-line work. This vital piece of equipment must be consistently tested and cared for to ensure that it is in proper operating condition.

3.4.1 Key Components and Features

Figure 16 shows some important components of the boom assembly. It is essential to note that the entire boom assembly is not insulated. Typically, the final, or upper, section of the boom is constructed of fiberglass and is the primary insulated portion. Some trucks may also be fitted with a fiberglass insert in a lower section. This would protect ground personnel in the event of a failure in the upper section. The collector shield (*Figure 17*) is a required

81301-12_F13.EPS

Figure 13 Ratcheting chain hoist.

81301-12_F15.EPS

Figure 15 Poly-dacron rope.

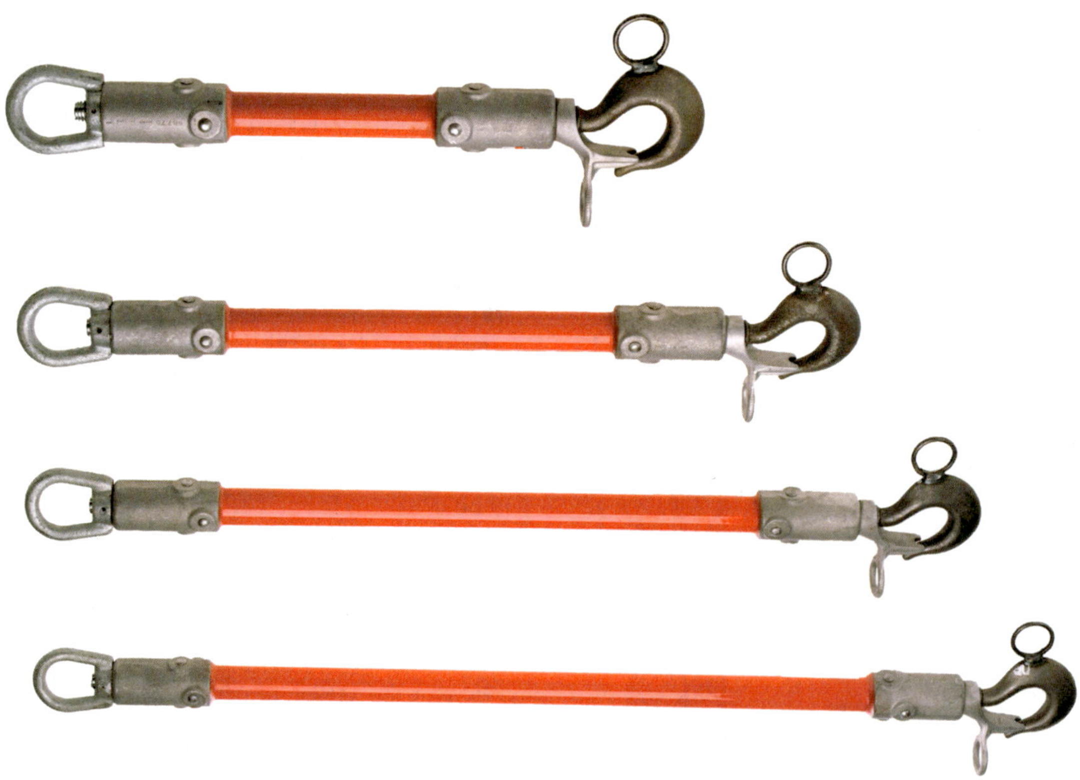

81301-12_F14.EPS

Figure 14 Strain link sticks.

Figure 16 Bucket truck components.

feature on truck booms that service 500kV and higher systems. The shield collects stray current and shunts it to ground. This significantly reduces current leakage through the boom at higher voltages. A corona shield (*Figure 18*) is used on all models to prevent corona from causing significant damage to the fiberglass portion of the boom.

The bucket is generally made of fiberglass and fitted with an aluminum liner (*Figure 19*). If there are two buckets, the aluminum liners must be electrically bonded together. The aluminum liner is then fitted with conductor-bonding cables and a bonding wand. The bonding cables may appear to be relatively short. Their length is intentionally limited for safety reasons, since a long, dangling cable could contact ground or another phase of power. The bonding cables must be equipped with spring clamps (*Figure 20*) for connection to the conductor. This prevents a forgotten connection from causing serious damage or injury if the bucket is moved while the clamp is left connected. A point of attachment (*Figure 21*) is provided for the worker's fall arrest lanyard.

The bucket truck is fitted with a permanent current leakage monitor (*Figure 22*). The moni-

Figure 17 Collector shield.

Figure 18 Corona shield.

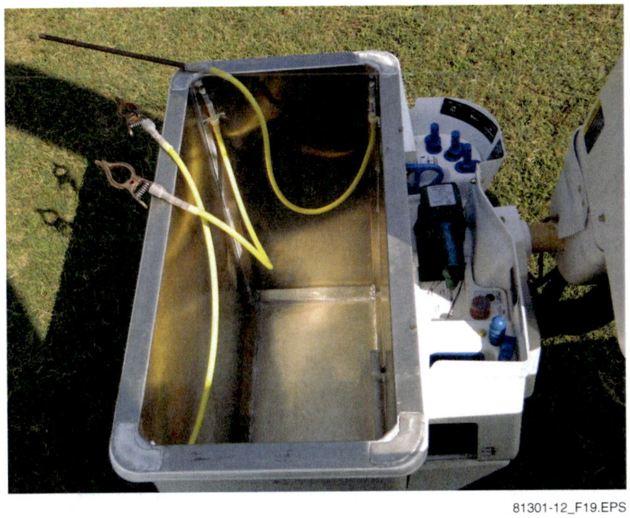

Figure 19 Aluminum bucket liner.

81301-12_F20.EPS

Figure 20 Bonding cable spring clamp.

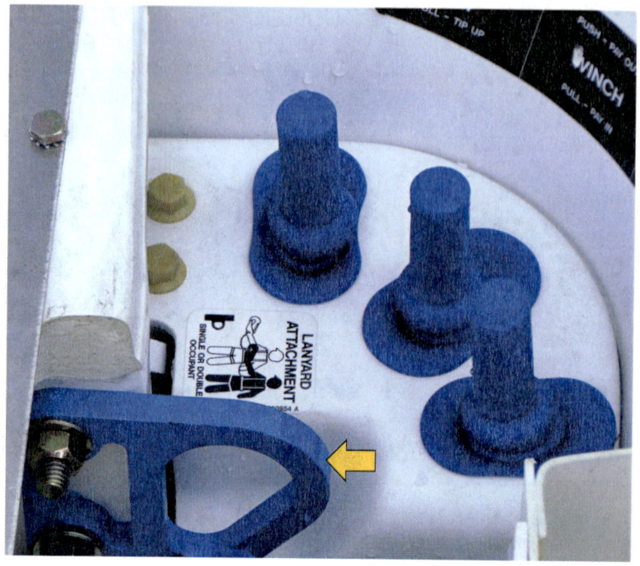

81301-12_F21.EPS

Figure 21 Fall arrest lanyard attachment point.

tor provides the current readings for testing purposes, and constantly monitors the leakage current while the truck is in use. Before use, the appropriate alarm setpoint for the present voltage is programmed. The alarm horn is contained within the monitor.

The hydraulic systems of bucket trucks are fitted with check valves in the lines to prevent a vacuum from forming if a hydraulic leak develops. This same hydraulic system can be used to power hydraulic tools. However, any hydraulic tools used should be considered a part of the truck itself and should not be swapped from one truck to another. Hydraulic lines for the power tools cannot be a part of, or connected to, the insulated portion of the boom. Many of today's current leakage monitor systems can separate the current readings of the insulated upper boom from the hydraulic lines and fluid.

3.4.2 Testing and Pre-Use Preparation

The bucket truck should be checked to ensure that the test sticker is affixed and shows that the annual inspection and testing is still valid. Place the bucket truck in the desired position on the job site. The outriggers are then extended and actuated. Most bucket trucks have a level (*Figure 23*) to ensure the outriggers are set to a position that results in a level truck.

Although some organizations may prefer a different order and have variations in the tasks, the following represents a typical truck preparation procedure.

- The truck must be grounded. Two temporary grounds are used. Grounds rods are driven into the soil and temporary grounding cables are attached. If the vehicle is not grounded, it should be barricaded and treated as if it is energized at all times.
- Before the boom is raised, it should be visually inspected and cleaned as necessary. The bucket must also be clean. Any type of accumulation on the boom or bucket can diminish their dielectric properties. A dirty surface tends to allow any water to remain in sheets, covering a larger surface area. Degreasers and solvents

On Site

Line Worker Accident

In 1980, a line worker trainee with Florida Power was killed while working from a bucket truck. A hydraulic leak developed in the boom hydraulic system, creating a vacuum in a portion of the boom hydraulic lines as the fluid leaked out. As a result, a flashover occurred between the truck and the conductor. An explosion destroyed the bucket and two workers fell roughly 75 feet. The trainee was killed while a second worker was left paralyzed.

The accident investigation did not reveal any sign of negligence. The incident was purely the result of a design flaw that allowed the vacuum to form. As a result of this incident, bucket truck hydraulic systems must be fitted with check valves to prevent a recurrence.

81301-12_F22.EPS

Figure 22 Current leakage monitor and alarm.

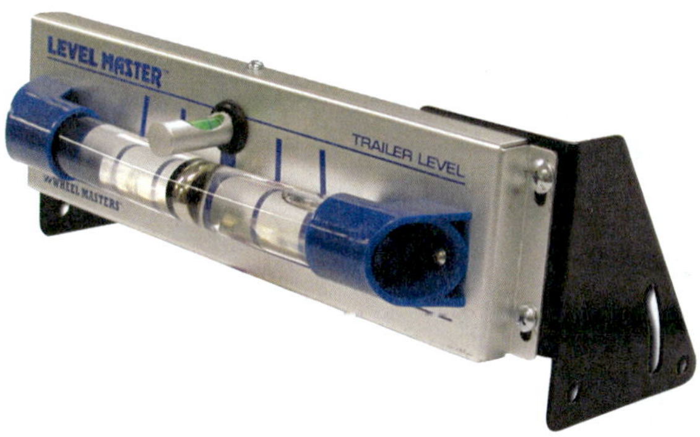

81301-12_F23.EPS

Figure 23 Vehicle level.

should not be used on fiberglass surfaces. Special cleaning agents designed for fiberglass surfaces should be used to protect the gel coat (*Figure 24*). These products are easily managed on the truck since they are individually packaged. A good coat of wax after cleaning also helps. The wax helps any water to separate into beads and drain off.

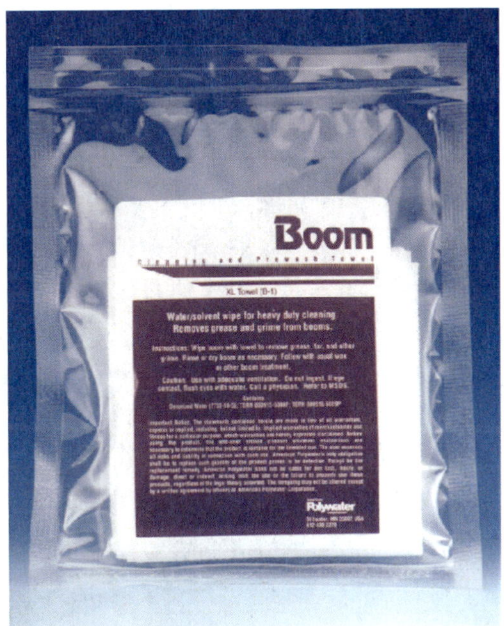

81301-12_F24.EPS

Figure 24 Boom cleaning wipes.

- Operate the boom and bucket through the range of motion from the lower controls first. Raise the bucket to its maximum height and leave it in this position for 5 minutes. This ensures that the hydraulic system is sound and can hold the bucket in position. Then operate the boom and bucket through the range of motion using the bucket controls.
- The boom should be checked to ensure that any measureable current leakage is within tolerance. OSHA requires that the boom be tested at the beginning of each work day, and whenever there is an increase in voltage of the lines being serviced. A record of every test must be kept and the results noted on the current leakage monitor. Additional information regarding the test, such as the weather conditions and temperature, are also important.
  - Refer to the current leakage monitor operating instructions. Current leakage monitors typically have a self-test function that should be initiated first. This will ensure that the meter is functioning properly. Most meters also allow for a continuity test of the truck-mounted circuitry. This ensures the circuits are intact. Then set the desired alarm level for the test. The actual test can now begin.
  - The bucket is raised without an occupant to a source energized at a voltage equal to what will be encountered. The bucket is positioned to make contact with the source. During this test, other workers standing on the ground should not be touching the truck. The boom

GOING GREEN

Transformer Insulating Medium Derived From Plants

Siemens has developed a new power transformer that uses plant oil as an insulating medium. The 40MVA capacity unit is being tested under real service conditions in the Black Forest of Germany.

81301-12_SA01.EPS

operator remains on the truck itself. The bucket must remain in contact for three minutes while the current leakage monitor is observed. Per *OSHA Standard 1910.269(q)(3)(xii)*, current leakage to ground may not exceed 1μA per kV of nominal phase-to-ground voltage. For example, 133kV is the typical phase-to-ground voltage for a 230kV line. Therefore, if the phase-to-ground voltage is 133kV, then 133μA would be the maximum allowed current flow. Work should be suspended immediately if this value is exceeded. The alarm setting of the monitor should be set at a value significantly lower than the maximum allowed current. Most employers require a setting that is roughly 50 percent of the maximum value. In this example, an alarm setting of 70μA would be common. This ensures that workers are notified long before the maximum value has been reached, assuming the failure is not sudden. Note that the current flow must never exceed 1 milliamp (mA) regardless of the voltage present. This is the same value as 1,000μA. If the current shows no significant deviations from other recorded tests and the value is within the 1μA/per kV phase-to-ground allowed, bare-hand work can proceed.

It is important to note that an operator must remain on the truck once work begins. However, operators are prohibited from using the bucket controls while a worker is in the bucket except during emergencies. The operator on the truck is also responsible for ensuring that other workers do not approach or make contact with the truck. Workers on the ground should also remain clear of the vehicle ground rods and cables while the truck is in operation and near energized lines. If the truck suddenly becomes energized, the step potential around the truck will be very high.

While work is being performed, the truck engine and hydraulic pumps should not be shut down. Monitoring these operations is also a responsibility of the truck operator.

4.0.0 Bare-Hand Work

Bare-hand work can be done from a bucket truck or similar aerial lift, a helicopter, or from the conductors themselves. When work is to be done from the conductors themselves, a line worker must first access the conductors safely. This can be done from a bucket or helicopter as well. Access to the conductors is also possible from a pole or tower by using insulated ladders of sufficient length.

Trainees are reminded that bare-hand workers must maintain the MAD between themselves and other phases of power or a grounded surface. An insulated measuring stick is required to be within the reach of bare-hand workers. The site supervisor is also responsible for ensuring that sufficient clearance exists for MADs to be maintained.

Throughout the procedures outlined here, regarding a worker making an initial connection to an energized conductor, it is assumed that the worker is wearing a conductive suit. For bare-hand work without a conductive suit, the procedures remain much the same with the exception of bonding the suit to a bucket or to a helicopter.

WARNING!

Placing the body or parts of the body between energized surfaces at different potentials, or between energized surfaces and any surface at ground potential, will result in serious injury or death.

4.1.0 Bucket Truck

OSHA requires that bucket trucks have a MAD table visible to the bucket occupant. When working from a bucket, workers must keep both legs inside the bucket at all times. At least one foot should always be on the floor. The fall arrest lanyard should be connected to the appropriate point provided on the bucket assembly. The conductive lanyard from the worker's suit is also tied off to the bucket.

In some cases, tools or materials may need to be lifted up to the worker. The bucket cannot be used as an anchor point for a hand line. However, the conductor can be used for this purpose.

Once the bucket is in position adjacent to an energized conductor, both the worker and bucket are already affected by induction. As a result, the potential difference in voltage between the conductor and the worker has already diminished to some degree. The worker reaches out with the tip of the bonding wand from the bucket (*Figure 25*). As the wand approaches the conductor, an arc is established. The length of the arc depends upon the difference in potential.

Once the wand makes contact with the conductor, the bucket, worker, and conductor reach equal potentials. While maintaining contact with the wand, the worker reaches out and connects the bonding cable spring clamp to the line. Now the wand can be stowed in the bucket. The bonding process is complete.

The bond is maintained throughout the work process. When the worker is ready to lower the

81301-12_F25.EPS

Figure 25 Bonding from a bucket.

bucket, the wand is again placed in contact with the line. The spring clamp is removed and placed inside the bucket. The wand is then drawn across the conductor to its tip, and the tip drawn away. The arc eventually breaks when the gap is sufficient between the two.

4.2.0 Insulated Ladder

Line workers can gain access to an energized conductor by using an insulated ladder. The ladder can be suspended from the conductor, and the worker is then lifted to the ladder from a bucket truck (*Figure 26*). This may be necessary when the transmission line is too high for the bucket to reach.

Workers can also use a ladder to access the line from a structure (*Figure 27*). In either case, the procedure to bond with the line is the same. The workers remaining on the structure, or cold end, are well outside the MAD.

When using a ladder, there is no bucket bonding cable to be connected. In this case, the line worker ties a conductive suit tether directly to the energized surface. The unused tether must be stowed away to prevent it from being entangled.

81301-12_F26.EPS

Figure 26 Accessing a line from a vertical ladder.

81301-12_F27.EPS

Figure 27 Accessing a line from a structure.

4.3.0 Helicopter

Workers can access transmission lines using a helicopter. The worker sits on a platform mounted to the helicopter, and a bonding wand is attached to the platform. In this approach, the worker and helicopter all must come to the same potential. As a result, the helicopter pilot must also wear a conductive suit. A wand is again used to make the initial contact with a conductor (*Figure 28*). If the line worker leaves the helicopter and remains on the line for later pick up, a bond must again be restored as the helicopter approaches (*Figure 29*).

Some tasks can be done while the worker remains on the platform. The installation of spacers and aircraft warning markers are examples of tasks that can be done in this manner (*Figure 30*). Some companies report that they can install a spacer on conductors in under one minute, including the time to fly from one spacer location to the next. Obviously this approach is only safe in fair weather with insignificant winds.

Another task that requires the line worker to remain on the helicopter is conductor splicing.

81301-12_F28.EPS

Figure 28 Conductor bonding from a helicopter.

81301-12_F29.EPS

Figure 29 Bonding with a returning helicopter.

81301-12_F30.EPS

Figure 30 Installing aircraft warning markers.

This is certainly a more challenging and lengthy task than installing spacers or markers. It requires great patience and skill on the part of both line worker and pilot. *Figure 31* shows the sequence of tasks required to make a splice while in flight. Note the conductive suit worn by the pilot, and the bonding cable connection from the aircraft.

The helicopter can also place the worker on the transmission lines or structure and then move away. To provide line workers with a safer alternative for lengthy tasks and those that require movement along the conductor, the helicopter may set both the worker and a cable cart on the transmission lines.

4.4.0 Cable Carts

Cable, or conductor, carts can be placed from a helicopter or by a bucket truck (*Figure 32*). The cart provides a means of movement between poles or towers. A close inspection of a transmission line can be made this way, and minor repairs can also be made. A complete splice cannot be made this way, due to the potential danger of a falling conductor.

Cable carts typically have four wheels, all of which bear some weight and are conductive. In addition to the wheels, safety chains or a similar component are also needed to prevent the cart from falling. Some are motor driven, while others are moved manually by the line worker (*Figure 33*). When a line worker is on the line to assist in positioning the cart, the wand should be used to bond with the approaching cart.

A. THE FIRST PULLING GRIP AND A CHAIN HOIST ARE PLACED.

B. A SECOND PULLING GRIP AND CABLES ARE CONNECTED.

C. THE DEFECTIVE SPLICE IS CUT OUT OF THE CONDUCTOR.

D. A HYDRAULIC PRESS IS USED TO CRIMP THE SPLICE.

E. LINEWORKER PREPARES TO RELEASE TENSION ON THE PULLING GRIPS.

81301-12_F31.EPS

Figure 31 Splicing a conductor in flight.

81301-12_F32.EPS

Figure 32 Placing a cable cart.

81301-12_F33.EPS

Figure 33 Using a cable cart.

SUMMARY

Working on energized power transmission systems is a necessity, since outages represent a significant hardship for both the consumer and the utility. Even with a de-energized line, the work is still challenging and not statistically safer. However, such work can be done in a safe and effective manner. In fact, live-line work is safely accomplished every day.

Bare-hand work brings the worker to the same electrical potential as the conductor or component being serviced. It is essential that MADs be maintained from surfaces at ground potential and other phases of power. The conductive suit worn for bare-hand work eliminates the flow of significant current through the worker's body by acting as a conduit for the current flow.

Line workers access energized surfaces in a variety of ways. Regardless of the method used, workers must bond with the energized surface. As the surface is approached, induction begins to raise the electrical potential of the worker prior to contact. As the initial contact is made, an arc is established. However, since the worker's electrical potential has already been raised, the energy of the arc is significantly reduced.

Conducting bare-hand work safely and efficiently requires training, practice, and confidence in the equipment and the line worker's skills. Bare-hand transmission line work is an exciting facet of the already exciting field of power transmission installation and service.

Review Questions

1. When an unforeseen problem is encountered during bare-hand procedures, _____.
   a. a work stoppage is required until the entire crew understands any changes in procedure
   b. the site supervisor discusses the problem with experienced workers while others continue
   c. the experienced workers continue while the supervisor uses it as a teachable moment for trainees
   d. cancellation of the project is required by OSHA until the issue has been reviewed and resolved by the company safety officer
2. The MAD must be adjusted for higher altitudes when working at an altitude above _____.
   a. 1,500 feet
   b. 3,000 feet
   c. 5,000 feet
   d. 6,500 feet
3. What is the ergonomic component that is added to the electrical component of a MAD for all voltages above 72.5kV?
   a. 1 foot
   b. 2 feet
   c. 3 feet
   d. 5 feet
4. Conductive footwear that meets ASTM testing standards is marked with the letters _____.
   a. CF
   b. HV
   c. CD
   d. EHV
5. Class C hard hats are tested in accordance with an ASTM standard to ensure that they are conductive.
   a. True
   b. False
6. The allowable current flow measurement through an insulated ladder during a daily field test is _____.
   a. 0.33 µA per kV phase-to-ground
   b. 0.33 mA per kV phase-to-ground
   c. 0.33 A per kV phase-to-ground
   d. 33 µA per kV phase-to-ground
7. A collector shield is required on bucket trucks that service systems at _____.
   a. 69kV and higher
   b. 115kV and higher
   c. 230kV and higher
   d. 500kV and higher
8. During a pre-use current leakage test, the bucket is raised to make contact with a power source energized at the same voltage as that to be serviced and held there for _____.
   a. 1 minute
   b. 3 minutes
   c. 5 minutes
   d. 10 minutes
9. Most employers require that the current leakage monitor alarm setting be roughly _____.
   a. 10 percent of the maximum value
   b. 25 percent of the maximum value
   c. 35 percent of the maximum value
   d. 50 percent of the maximum value
10. When a line worker accesses an energized conductor using a ladder instead of a bucket truck, _____.
    a. no conductive tether needs to be connected
    b. both conductive suit tethers must be tied to the energized surface
    c. the two conductive suit tethers must be tied together and stowed
    d. one conductive suit tether is tied to the energized surface; the other is stowed

Trade Terms Introduced in This Module

Ergonomic component: A distance value that is added to the electrical component of the minimum approach distance to reach the MAD values shown in *OSHA Standard 1910.269(l)(10), Tables R-6 through R-9*. For voltages up to 72.5kV, the ergonomic component is 2 feet; for voltages above 72.5kV, the ergonomic component is 1 foot.

Hold off: Term describing the request by linemen for any automatic reclosers related to the section of line they are working on to be disabled or set to a non-automatic position; also known as a one-shot request.

Additional Resources

This module presents thorough resources for task training. The following resource material is suggested further study.

OSHA Publication 3071, Job Hazard Analysis Guide.

Live Line Work Practices, Second Edition. Alexander Publications.

Figure Credits

Courtesy of Juan Guevara, JRG Photography, Module opener

KT Power, Inc., Figures 1, 10, 25–27, 32, and 33

US Department of Labor, Occupational Safety & Health Administration, Table 1

Hubbell Power Systems, Figures 4, 7–9, 12 (top photo), and 13–15

Courtesy of Warson Brands, Figure 5

Siemens Corporation, Figure 6 and SA01

Hastings, Figure 11

Images provided courtesy of Haverfield International Incorporated, Figures 12 (bottom photo) and 28–31

Topaz Publications, Inc., Figures 16–22

Courtesy of Wheel Masters, Figure 23

Courtesy of American Polywater Corporation, Figure 24

NCCER CURRICULA — USER UPDATE

NCCER makes every effort to keep its textbooks up-to-date and free of technical errors. We appreciate your help in this process. If you find an error, a typographical mistake, or an inaccuracy in NCCER's curricula, please fill out this form (or a photocopy), or complete the online form at **www.nccer.org/olf**. Be sure to include the exact module ID number, page number, a detailed description, and your recommended correction. Your input will be brought to the attention of the Authoring Team. Thank you for your assistance.

Instructors – If you have an idea for improving this textbook, or have found that additional materials were necessary to teach this module effectively, please let us know so that we may present your suggestions to the Authoring Team.

NCCER Product Development and Revision
13614 Progress Blvd., Alachua, FL 32615
Email: curriculum@nccer.org
Online: www.nccer.org/olf

❑ Trainee Guide ❑ AIG ❑ Exam ❑ PowerPoints Other ______________________________

Craft / Level: Copyright Date:

Module ID Number / Title:

Section Number(s):

Description:

Recommended Correction:

Your Name:

Address:

Email: Phone:

81302-12

Reconductoring Transmission Lines

Module Two

Trainees with successful module completions may be eligible for credentialing through NCCER's National Registry. To learn more, go to **www.nccer.org** or contact us at **1.888.622.3720**. Our website has information on the latest product releases and training, as well as online versions of our *Cornerstone* newsletter and Pearson's product catalog.

Your feedback is welcome. You may email your comments to **curriculum@nccer.org**, send general comments and inquiries to **info@nccer.org**, or fill in the User Update form at the back of this module.

81302-12
Reconductoring Transmission Lines

Objectives

When you have completed this module, you will be able to do the following:

1. Plan a reconductoring project, including:
   - Obtaining permits
   - Selecting equipment and materials
   - Establishing the starting point
2. Prepare for a reconductoring project, including:
   - Performing a pre-job briefing
   - Equipment setup
   - Setting up guard structures
3. Describe how to pull conductors by alleying-out existing conductors.
4. Describe how to pull conductors using the existing conductors.

Performance Tasks

Under the supervision of your instructor, you should be able to do the following:

1. Develop a pulling plan.
2. Properly fill out a pre-job briefing.
3. Properly set up and use a tensioner and puller.
4. Alley-out wire and either install a layout structure or pull wire using the old conductor.

Trade Term

Sheave

Industry Recognized Credentials

If you're training through an NCCER-accredited sponsor you may be eligible for credentials from NCCER's Registry. The ID number for this module is 81302-12. Note that this module may have been used in other NCCER curricula and may apply to other level completions. Contact NCCER's Registry at 888.622.3720 or go to **nccer.org** for more information.

Contents

Topics to be presented in this module include:

Figures and Tables

1.0.0 Introduction

An earlier module covered the basics of transmission line installation. While this module focuses on replacing existing conductors, it also provides more advanced information about conductoring methods and equipment. Reconductoring may be done to replace damaged or worn-out conductors or to increase the capacity of the system. Effective planning is the key to an efficient installation. Before any work can start, the amount and types of equipment and materials needed for the job must be determined. This includes calculating the amount of wire that is needed. If the installation crosses private land, federal land, highways, or waterways, the necessary permits must be obtained.

Once the planning is complete and the materials and equipment are ordered, the installation work site must be laid out and guard structures must be built to protect people and property from falling lines. Rights-of-way may need to be cleared as well. It may also be necessary to reinforce the existing transmission structures to account for the increased weight of conductors and insulators. Before a single conductor is pulled, insulators may need to be changed-out and travelers hung from the insulators (*Figure 1*).

NOTE

The term *traveler* is used in this module. *Stringing block* and *roller* are other terms commonly used in the industry.

Safety of work crews and the general public is of paramount importance. In populated areas in may be necessary to have security guards and safety watchers. Crews need to be protected from induced voltages from nearby lines, especially when the reconductoring is done while the original lines remain energized.

81302-12_F01.EPS

Figure 1 Travelers attached to insulators.

1.1.0 Reconductoring Methods

Reconductoring, by definition, implies that existing conductors are being replaced. In some cases, it is possible to reconfigure the power grid so that the affected substation can be temporarily supplied from another source. In that case, the affected transmission line can be shut down and the existing conductors can be used to pull the new conductors. If the service cannot be taken off-line, the new conductors must be installed while the existing energized conductors are disconnected from their insulators and pulled out of the way. This increases the level of hazard and requires special attention to safety.

There are a number of reconductoring scenarios, which are summarized as follows:

- In some cases, the existing structure is to be retained, but because of the weight of the new conductors, it must be reinforced before the new conductors are pulled.
- Sometimes, the entire structure is replaced. This can happen simply because of the age of the structure or because of a need to increase the capacity of the line. In such cases, the new structure is erected, then either the existing conductors are transferred or new conductors are pulled. In these cases, the insulators may need to be replaced to accommodate a higher voltage.
- It is sometimes necessary to replace existing conductors without interrupting the power for an extended period. In those cases, the existing conductors are disconnected from their insulators and alleyed-out while the new conductors are installed.
- When it is possible to de-energize the line while replacing conductors, the existing conductors can be used in place of a bull rope to pull the new conductors. The existing conductor is wound onto a takeup reel as the new conductor is pulled.

The tensioning method is commonly used when pulling in new conductors. With this method, the conductors are kept under tension as they are pulled, so they do not touch the ground. Contact with the ground can damage the conductors. Tensioning the conductors also helps to control them, so they are less likely to come into contact with live conductors as they are being pulled. Tensioning also keeps the conductors elevated when crossing highways or bodies of water.

There are situations in which the conductor is deposited on the ground. This is known as the slack or layout method. The conductor reel is placed on a jack stand in the bed of a truck in a way that allows it to freely rotate. The conductor is secured at one end. As the truck moves forward, the conductor is deposited on the ground, and then lifted to the insulators on the structure.

2.0.0 Planning

The reconductoring of transmission lines is a big job. Extensive planning is needed to ensure that the work is done safely and efficiently. Most companies require a written plan to be prepared. A critical element of that plan is determining the material, equipment, and personnel needed for the job. The *Appendix* contains an example of a form used in planning a wire-stringing project. A key element of this planning is identification of the risk factors and a plan to mitigate them.

A transmission line runs for many miles. It is common to break down the reconductoring into sections of one to four miles. This is not an arbitrary decision. Part of this determination involves identifying locations where the terrain is suitable for setting up pulling and tensioning sites. A number of factors enter into the selection of pulling sites, including the following:

- Safety of the public
- Accessibility of the sites
- Conductor length
- Capacity of the pulling equipment
- Location of dead ends

The job foreman or supervisor reviews the drawings for the section to determine the number and types of transmission structures, the height of each structure, and the distance between structures. Using this information, the amount of wire needed for a section can be determined. This information can also be used to determine the number and types of stringing equipment and accessories, such as the following:

- Stringing blocks (travelers)
- Insulators (if required)
- Running grounds
- Running boards (if required)
- Dead ends
- Armor rods
- Spacers
- Vibration isolators
- Pilot line rope
- Bull rope
- Hard lines (wire rope used for long pulls and bundled conductors)
- Finger line (P-line)
- Kellems grips, bands, banding tool, and tape
- Tensioner
- Pilot line puller
- Bullwheel puller
- Takeup reels
- Conductor reels

When calculating the amount of wire needed for a job, it is common to add the distances between structures to figure the total length of the run, and then to add 10 percent to account for jumpers and line sag. This total would then be multiplied by the number of conductors to be run.

NOTE

Sag requirements must be determined by the cognizant engineers. The planners must provide the engineers with required data well in advance of stringing operations so that they have time to calculate the sag.

Depending on the terrain, it may be necessary to use helicopters for part of the preparation and stringing. This also needs to be planned in advance.

2.1.0 Permits

Transmission lines often cross highways and waterways. They may pass through easements on private property; federal, state, or municipal land or parks; and areas that are protected because of wildlife habitat. There is no shortage of regulations governing access to these areas. In many cases, permits are required before work can begin. In some cases, the company may have to file an environmental impact statement.

While it may not be up to the crew supervisor to obtain permits, it is important that he or she make sure that all the necessary permits have been requested and issued. Crew supervisors must also be aware of the rules and regulations governing access to, and use of, protected areas, as well as regulations governing pollution of land, waterways, and storm water systems. This topic is covered in some detail in the module *Transmission System Maintenance.*

Among the agencies that may require permits for work on transmission lines are the following:

- *US Department of Transportation* – The Federal Highway Administration issues permits for utility projects that cross highways that receive federal aid. The state DOT may require a permit for projects that cross state-maintained highways or use rights-of-way along those highways.

On Site

Hands-On Planning

Those involved in planning the reconductoring operation should travel the site to look for obstructions, determine if clearing is needed, and scout locations for setup of pulling and tensioning equipment. They should also check the condition of existing transmission structures to see if any pre-project maintenance needs to be done. On a map or drawing, the site is likely to seem clear, dry, and unobstructed. The reality may be far different. Snow melt and spring rains, for example, may make a site difficult to access. A location that seems ideal as a place to set up for pulling wire may even be underwater.

- *US Army Corps of Engineers* – The Corps of Engineers regulates construction or work performed in navigable waterways.
- *US Department of Agriculture* – The National Forest Service grants authorization for use of national forest land.
- *US Department of Interior* – The National Park Service issues special-use permits for utility systems and facilities on national park land.
- *US Environmental Protection Agency* – Under the Clean Water Act, the EPA, in coordination with states, governs release of pollutants into waterways.
- *US Fish and Wildlife Service* – The US Fish and Wildlife Service (FWS) regulates activities that may affect migratory birds as well as threatened and endangered species of birds, animals, and aquatic life. FWS would issue permits for and regulate work in wildlife refuges and bird sanctuaries.

Local governments may also need to be consulted. If any of the lines cross railroad tracks, the railroad must also get involved.

GOING GREEN

Recycling Conductors

When existing transmission lines are replaced, there can be miles of removed conductor. This material can be recycled to recover the steel, copper, and aluminum.

3.0.0 Preparation

A considerable amount of work needs to be done before stringing can begin. The first and foremost of the preparatory activities is to make sure that all safety concerns have been considered. Other key tasks are setup of the pulling and tensioning sites and preparation of the transmission structures. Of utmost importance is the protection of the public in populated areas. This may require fencing-in the pulling and tensioning sites and posting security guards at areas accessible to the public. In urban areas or areas with high-induction potential (induction corridors), barricades must be used to protect the public.

3.1.0 Job-Site Hazards

One of the key concerns for workers stringing wire is the hazard from induced voltages. Transmission lines often share a right-of-way (*Figure 2*) or cross each other. Even when the line being replaced is de-energized, nearby high-voltage lines can induce a deadly voltage into that line. A 500kV line might be capable of inducing 30kV into the de-energized line, for example. Installers must, at all times, remain outside the minimum approach distance (MAD) for the voltage being carried by nearby lines. This requirement applies to phases of the transmission lines being reconductored if the work is being done with power applied to those phases. All equipment used in the process, as well as structures and conductors, must be bonded together.

WARNING!

When applying grounds, always connect to the ground side first, then connect to the equipment or structure. Do the opposite when disconnecting the ground.

The possibility of induced voltages also creates a need to ground equipment such as trucks and trailers used for tensioning, pulling new conductors, and recovering old conductors. In cases where bystanders could have access to the equipment, it may be necessary to build a fenced enclosure and bond it to the ground grid.

Other electrical hazards are present at the job site. These hazards include the following:

- Faults caused by accidental contact with energized lines
- Lightning strikes
- Accidental energizing of de-energized lines
- Static charges caused by environmental conditions

In order to protect workers, it is critical that equipotential zones be set up in work areas.

3.2.0 Guard Structures

In populated areas, where transmission lines have the potential to fall on people, buildings, highways, or distribution lines, guard structures must be built. These structures prevent conductors from falling onto pedestrians or roadways. A typical support structure consists of two wood uprights with a cross-member, although some can be more elaborate. *Figure 3A* shows guard structures set up to protect power distribution lines. In *Figure 3B*, two cranes are used to support a beam that is serving as a guard structure to protect another high-voltage transmission line. Guard structures are commonly placed on both sides of intersections, roads, walkways, or rail lines that are crossed by the transmission lines. These structures are removed when the reconductoring is finished. In addition to guard structures, netting may be required in some areas.

3.3.0 Working With Helicopters

Helicopters are often used to install pulling ropes, especially in remote areas where access is difficult for ground traffic. In such cases, some special safety issues must be considered. Operating with helicopters, though routine, is dangerous. When flying next to power lines, factors such as winds, workers outside the aircraft, external loads and weather can conspire to create hazards. Thorough work planning that clearly outlines job tasks in step-by-step detail, listing individual responsibilities, emergency procedures, and a weather briefing ensures helicopter work remains safe. Refer often to *FAA Regulations Part 133* and *Advisory Circular AC 133-1A – Rotorcraft External-Load Operations* for detailed regulations for helicopter operations.

3.3.1 Helicopter-Specific PPE

Line workers wear PPE when working with helicopters, as with any other maintenance task, but rotor downwash adds requirements. Eye protection is necessary to protect from flying debris. Hard hats must have a chinstrap. Many helicopter-borne workers wear an aviator cranial helmet with chinstrap fastened, goggles in place, and communication with the pilot. Workers around helicopters must wear tight-fitting clothes; flapping clothing can get caught in the hoist line or snag on uneven,

81302-12_F02.EPS

Figure 2 Parallel transmission lines.

(A)

(B)

81302-12_F03.EPS

Figure 3 Guard structures.

jagged tower surfaces. A fall-arrest system must be used when standing on the helicopter skid or working from a platform outside the helicopter. Tools should be on lanyards to prevent them from falling and injuring someone on the ground. If voice communication between the workers and the pilot is not available, hand signals are used to ensure effective communication.

Line workers often stand or sit on the landing skids or a work platform while performing maintenance. Working from the side of a helicopter places the workers in the downwash from the rotors and they are subjected to loud noise and cold. Dressing warmly with good hearing protection is essential. In cool weather, workers may need to wear exposure suits.

3.3.2 Helicopters on the Ground

Workers must approach helicopters from the side or the front, but always within the range of the pilot's vision. Spinning rotors can sag at low speed, so they are not always the same height above the ground. Walk crouched over when under the rotor, even when it appears well above the head. Stay far away from the tail rotor. If suddenly blinded by dirt, sit down and wait for assistance. Never try to grope or feel the way to or from the helicopter.

If the takeoff site is on a hill, never approach or depart the helicopter on the upslope side. Also, never enter the area of lowest rotor clearance. To avoid contact with the main rotor blade, carry long pieces of equipment or tools at or below the waist. Never carry long equipment or tools upright or over the shoulder.

The downwash area can be as much as 100 feet from a helicopter on the ground. Do not set any loose equipment in the downwash area. A seemingly harmless object can become a dangerous unguided missile, particularly from the strong downwash during a takeoff. Helicopter engines are subject to foreign object damage from a small object being sucked into the engine intake.

3.3.3 Helicopter Hoisting Operations

Helicopter hoisting operations are inherently dangerous. All ground personnel must be trained in the proper procedures. They should know the hand signals used during an external load operation. There should be a landing signal operator (LSO) in a position visible to the pilot and the hookup person. The LSO stands in front of the aircraft and to the right of the flight path and as close to the operating area as possible with his/her back to the wind.

WARNING!

Do not hook or unhook a load unless properly trained. High levels of static electricity can build up in loads suspended beneath helicopters. Never touch an ungrounded suspended load. Ground loads by touching the hoist cable using a grounded insulated tool with a pull-away clamp. If possible, the LSO will direct the pilot to set the load on the ground in order to discharge static electricity before anyone touches the load.

After hooking up the cargo sling, ground personnel must move clear of the aircraft. This prevents load cables from entangling a ground crew member and hoisting him or her aloft. Staying well clear also ensures that a swinging load does not strike a ground worker. After workers are clear, the LSO can signal the pilot to take off.

In some cases, workers are in a basket suspended well below the helicopter. All baskets and hoist cables must be inspected in accordance with *FAA Part 133*.

3.3.4 Fueling Operations

Fueling a helicopter in the field has become a routine operation. However, it is routine only because workers know and follow safety precautions. All components of the refueling process must be grounded to the same potential to prevent sparks from static electricity.

- Establish a good ground by driving a conductive rod into the ground.
- Ground the fuel tank and fuel pump to the ground rod.
- Ground the helicopter to the ground rod.
- Touch the nozzle to the helicopter before inserting it into the fuel inlet. Approved fuel hoses have an internal ground bonded to the fuel pump.

WARNING!

Refueling should be stopped immediately if there is lightning in the area. Also, do not smoke within 100 feet of refueling operations.

3.4.0 Installing Travelers

Travelers are available in a variety of configurations, ranging from single-line travelers to bundled conductor travelers designed for stringing up to four conductors at the same time. These travelers are specially constructed for the purpose, with one **sheave** for each conductor and a separate sheave for the pulling rope. *Figure 4* shows examples of single-sheave and multi-sheave travelers. Some travelers are equipped with a side gate that allows pulling lines or conductors to be placed directly over the sheave (*Figure 5*).

Special travelers are used when stringing is done from a helicopter. These travelers have outrigger arms (*Figure 6*) that direct the pilot line into the traveler. A spring-loaded gate secures the line. The outrigger is usually painted a bright color so it can easily be recognized from the air.

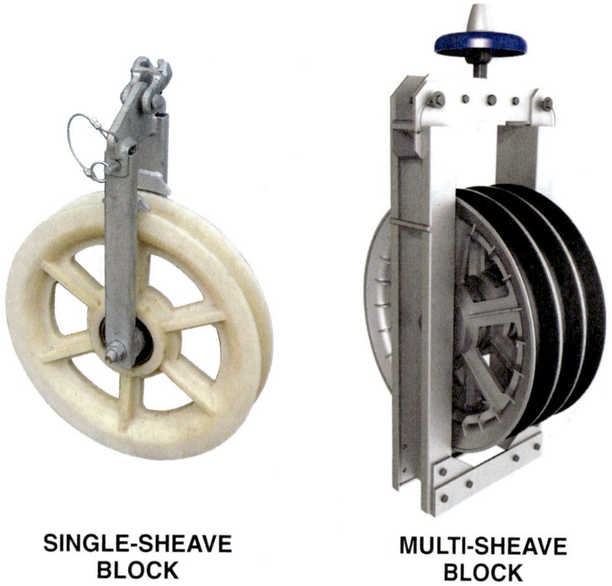

81302-12_F04.EPS

Figure 4 Examples of travelers.

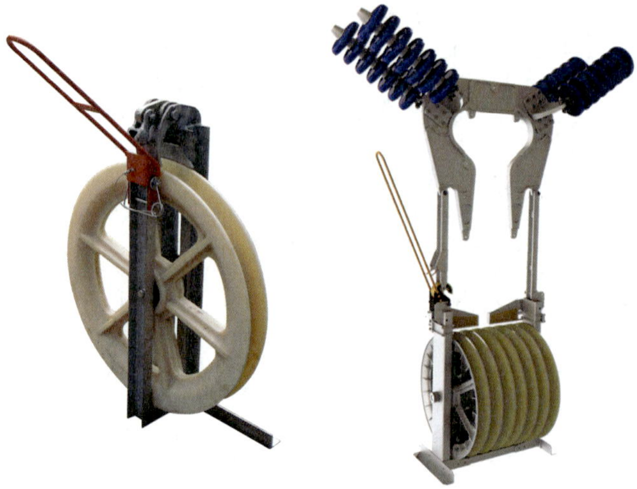

81302-12_F06.EPS

Figure 6 Traveler with an outrigger arm.

Travelers are attached to the insulators. The attachment method depends on the insulator configuration. When single insulators are used, travelers are hung directly from the conductor shoe. For V-type insulators, travelers are attached to the yoke plate. Depending on the level of hazard, it may be necessary to ground the traveler. The traveler itself does not provide a ground and can be damaged by stray currents. Grounding attachments that can be connected to a ground source are available for this purpose (*Figure 7*).

81302-12_F05.EPS

Figure 5 Side gate open.

81302-12_F07.EPS

Figure 7 Traveler with grounding attachment.

> **NOTE**
> If grounding attachments are used with the travelers, the attachments are normally positioned so that they point toward the pulling end.

If the insulators are being replaced, the travelers are often installed before the insulators are raised into place. At the same time, it is common to place a finger line (also known as a P-line) over the traveler and then lower the line to the ground and tie it off (*Figure 8*). This line is then used to pull the pilot line from one structure to the next, where it is connected to the finger line for that structure. The pilot line is pulled over the traveler, down to the ground, then walked to the next structure where the process is repeated. This is not the only method used, however. The pilot line can also be placed by workers in bucket trucks when they hang the insulators (*Figure 9*). Helicopters are also used to string pilot lines.

3.5.0 The Pulling Equipment

Pulling equipment is set up near the dead end at each end of the span to be pulled. Before the pulling equipment can be set up, a temporary ground grid must be installed at each end of the pull. The machines used to do the pulling are then positioned on the ground grid. This provides protection for workers and equipment from induced voltages and fault currents.

The amount and type of pulling equipment depends on the type of pull being performed. For example, the equipment needed to pull conductors on newly installed transmission structures is the same as the equipment needed to pull conductors while the existing lines remain energized. In addition to a bullwheel puller and tensioner, a pilot line winder would be required. When the transmission lines can be de-energized, the existing conductors can be used to pull the new conductors. Therefore, the pilot line, bull rope, and related equipment are not needed.

The pilot line winder (*Figure 10*) is a powered winch that is used to feed out and recover the pilot line. The pilot line is connected to the bull line, which is wound onto the reel(s) on the bullwheel puller (*Figure 11*). The bullwheel puller is equipped with a level-wind arm that keeps the line from bunching up on the reel. The bullwheel puller is positioned at the opposite end of the conductor run. Once the pilot line reaches the bullwheel puller, the two lines are connected together. The pilot line is then retracted, pulling the bull rope through the travelers as it progresses. The bull rope is usually kept under tension using a tensioner (*Figure 12*).

81302-12_F08.EPS

Figure 8 Finger lines installed on travelers.

81302-12_F09.EPS

Figure 9 Installing travelers from a bucket truck.

81302-12_F10.EPS

Figure 10 Pilot line winder.

81302-12_F11.EPS

Figure 11 Bullwheel puller.

The tensioner is equipped with a tensioning wheel, which the tensioner operator controls in order to maintain the required tension on the conductor. The conductor is wrapped around the wheel and placed in the groove on the wheel.

Once the end of the bull rope has reached the puller, the conductor reel is set up, and the conductor is attached to the bull rope using a Kellems grip (*Figure 13A*), which is then taped and banded to keep it from catching on the travelers. A swivel is used to connect the Kellems grip to the bull rope (*Figure 13B*). This prevents the bull rope from twisting as it is pulled through the travelers. The bullwheel puller on the far end is then used to pull the conductor(s) through the travelers, while the conductors are fed through the tensioner on the other end (*Figure 14*).

Before the pull begins, a running ground (*Figure 15*) must be secured to the conductor and attached to a suitable ground. The running ground bleeds off static electricity that can build up during the pull. Once the pull is underway, the rope attached to the running ground can be secured for the remainder of the pull.

81302-12_F12.EPS

Figure 12 Tensioner.

NOTE

If more than one length of conductor is required for the pull, it is common practice to connect the two lengths together with a Kellems grip for the pull. Once the pull is complete, the conductors can be lowered to the ground and spliced, or the splice can be performed from an aerial bucket.

A. KELLEMS GRIP

B. SWIVEL

81302-12_F13.EPS

Figure 13 Using a Kellems grip.

81302-12_F14.EPS

Figure 14 Conductor reels and tensioner.

At the conclusion of the pull, one end of the conductor is attached to the insulator at one end of the line so the line can be sagged. The sagging process was described in another module. Once the sagging is complete, the other end of the conductor is permanently secured to its insulator string. At that point, armor rods are applied (*Figure 16*) and the conductors are clipped-in to their insulators. Vibration isolators (*Figure 17*) are added and spacers are applied to bundled conductors as determined by the engineering plans.

On Site

Line-Throwing Gun

The Bridger™ throwing gun is designed to throw a line a great distance. In utility work, it is used to breach obstacles such as canyons, rivers, and other bodies of water.

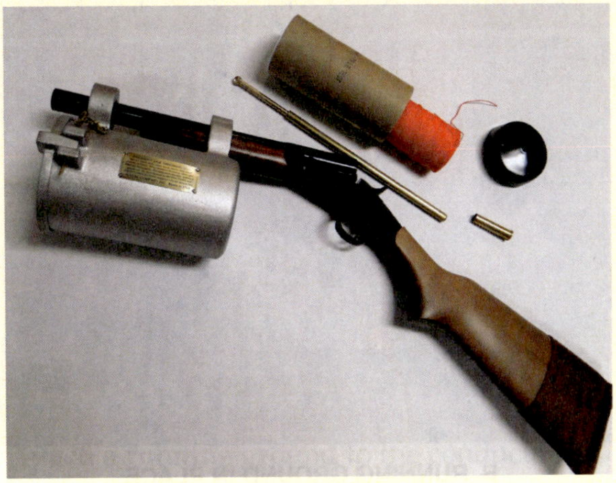

81302-12_SA02.EPS

81302-12_F19.EPS

Figure 19 Supporting conductors.

81302-12_F20.EPS

Figure 20 Conductor pull in process.

On Site

Deadly Static Electricity

During installation, conductors can pick up a static charge or may pick up induced voltage from nearby conductors. For this reason, a running ground must always be installed on the conductors during a conductor pull.

81302-12_F21.EPS

Figure 21 Crane supporting conductors.

MasTec

WIRE STRINGING RISK ASSESSMENT

Do not leave any blanks, enter "NA" if not applicable to this pull.

Foreman's Name: ____________________ Job #: ______________ Date: ____________

Customer Name: ______________________________ Work Order #: ____________

Number of wire stringing operations required on project: Work Location: ____________

Pull number currently being performed

1st	2nd	3rd	4th	5th

Length of pull: ______________

Wire size being pulled: ______________

Single or bundle conductor: ______________

Weight of conductor per foot: ______________

Ruling span: ______________

PRE-PULL HAZARD ASSESSMENT

Tooling Assessment for Conductor Pull

What is the safe working load of the travelers (dollies)? (to include angles, take off, and dead ends) ______________

What is the minimum diameter and type of traveler to be used on the angles and dead-ends? ______________

Indicate the rope size or conductor size being used and its load rating: ______________

What is the initial sag tension for this pull of conductor? ______________

If a dynamometer is being utilized for sagging purposes, is the calibration current? ______________

What size of hoist will be utilized? ______________

What is the safe working load of the slings that will be used? ______________

What type and size of Kellum grip will be used? ______________

What type and size of conductor/ohgw grip will be used? ______________

What is the safe working load of the swivel to be used? ______________

Are the dead-ends mechanical or compression? ______________

Has the press head and motor been inspected to ensure safe & reliable operation? ______________

What size of dies and press are required for the pull of conductor? ______________

Have the hydraulic and/or ratchet cutters been inspected to ensure safe & reliable operation? ______________

ROPE & WIRE INSPECTION PRIOR TO PULLING WIRE

What is the diameter and safe working load for the following:	Dia.	Safe Working Load
The rope being utilized to pull in the conductor		
The handline being utilized to pull in the conductor		
The existing conductor being utilized to pull in the conductor/ Reduce old conductor tensile strength by at least 25%		

	Yes	No	NA
Will the rope be inspected as it is payed out?			
Has the rope been preload tested at the calculated tension before each pull?			
If an ARD (anti rotational device) is being utilized, has it been inspected for safe & reliable operation?			
Are there any splices in the rope and if so, have they been inspected to ensure reliability			
Has the eye at the end of the rope been inspected to ensure reliability?			
If existing conductor is being utilized to pull in the new conductor, has it been inspected?			
If existing line sleeves are located, will they be cut out and kellums installed?			
Have the kellums been banded?			

Page **1** of **3**

81302-12_A01.EPS

EQUIPMENT INSPECTION

Puller make and Model: ____________________

Puller's maximum rated capacity: ____________________

Tensioner make and model: ____________________

	Yes	No	NA
Do the gauges on puller & tensioner work properly?			
Has the reel stand been inspected for proper braking?			
Has the tensioner been set up properly with regards to the take-off structure? (alignment)			
Have the puller, tensioner, reel stand and/or wire boat been anchored to ensure zero movement?			
Are traveling grounds being utilized at the tensioner?			
Are grounded travelers being utilized at energized crossings?			
If a V-groove or Wire Winder is being utilized, are traveling grounds being utilized?			
Are grounding mats being utilized at both the tensioner, reel stand and puller?			
Are the tensioner, puller and reel stand grounded and/or barricaded?			
If mobile equipment is being utilized as a guard structure, has it been grounded and/or barricaded?			
Are EPZ grounding methods being utilized if required?			

WIRE PULLING OPERATIONS

What is the planned conductor sag (in feet) during pulling operations?	
Calculate and indicate your required pulling tension (based on ruling span, line pull & conductor weight)	
What is the max gauge pressure required on tensioner for this pull?	
What will the operator set the initial gauge pressure on the tensioner to begin this pull?	
What is the maximum gauge pressure required on the puller for this pull?	
Longest span involved in pull:	
Indicate pull line being used for this pull (rope, existing conductor, or hardline):	
What is the safe working load of the travelers, to include angles, take-off and dead-end?	
If pulling through any dead-ends or angles, what is the minimum diameter of traveler to be utilized?	

	List "Red Zone Area" (Transmission or Distribution Line Crossings, R/R, Interstates or Road Crossings, Buildings, Parking Lots, etc.)	List "Type of Guarding" required for each area
1		
2		
3		
4		
5		
6		
7		
8		
9		
10		

Page **2** of **3**

81302-12_A02.EPS

How many men do you have on your crew?	
How many men will be required at the puller for pulling and wire catch-off?	
How many men will be required at the tensioner for tensioning and wire catch-off?	
How many "Red Zone" locations will require personnel?	

	Yes	No	NA
Will you have radio communication for each "Red Zone" location?			
Do you have the adequate number of employees to monitor each red zone area?			
If pulling over existing transmission or distribution lines, have they been requested to be de-energized?			
If pulling over transmission or distribution lines that CANNOT be de-energized, has a NRO (One Shot) been issued?			

NOTE: Reliable communications is required at both puller and tensioner throughout the pull

	Yes	No	NA
Has a copy of the final "Wire Stringing Risk Assessment" been given to your inspector?			
Has this wire stringing Risk Assessment information been communicated to the crew members?			

	Low	Medium	High
Is this considered a low, medium, or high risk wire stringing procedure?			

- "Low Risk" — Complete the form and review by General Foreman and/or Safety Specialist
- "Medium Risk" — Complete form and review by Safety Specialist and/or General Foreman and/or Manager
- "High Risk" — Complete form and review by Safety Specialist and General Foreman and/or Manager
- "High Risk" — Safety Specialist and/or General Foreman must be present during pulling procedure.

Foreman's Signature	Date	General Foreman's Signature	Date
Safety Specialist's Signature	Date	Division Manager's Signature	Date

KEEP A COPY OF "WIRE STRINGING RISK ASSESSMENT" FORM ON JOB AT ALL TIMES

Page **3** of **3**

81302-12_A03.EPS

Additional Resources

This module presents thorough resources for task training. The following resource material is suggested further study.

OSHA Publication 3071, Job Hazard Analysis Guide.

Live Line Work Practices, Second Edition. Alexander Publications.

Figure Credits

Terry Williams, Module opener, Figures 1, 9, 12, and SA01

KT Power, Inc., Figures 2, 16, 19, 21, and 22

Topaz Publications, Inc., Figures 3, 5, 8, 11, 13, 14, 15 (bottom photo), 17, 20, and SA02

Courtesy of TSE International, Figure 4 (left photo), 6 (left photo), and 10

Courtesy of Sherman & Reilly, Inc., Figure 4 (right photo), 6 (right photo), 7, and 18

Courtesy of SAFETYTEST, Figure 15 (top photo)

Courtesy of Mas Tec, Inc., Appendix

81303-12

Construction Maintenance and Repair – Hot Stick

Module Three

Contents

Topics to be presented in this module include:

Figures and Tables

Figures and Tables (*continued*)

1.0.0 Introduction

There are few tasks that allow transmission lines to be de-energized for service, repair, or even conductor replacement. Nearly every task requires line workers to work on energized circuits. The interruption of electrical service to consumers is a serious matter, so working on circuits while they remain live is an essential and common occurrence. Of course, a great deal of de-energized construction must be done first to create the transmission system (*Figure 1*).

Although it would seem that live-line work is far more hazardous than working with de-energized circuits, it is not necessarily the case. Live-line work is completed safely on a daily basis, and there is no reason why that cannot remain true. Workers are injured working on circuits thought to be de-energized that are actually live or become energized during the project. Mistakes in circuit identification can also lead to injury. In de-energized work, a worker often prepares for work near live circuits, but not for working directly with live circuits. In live-line work, all preparations and plans are made under the assumption that everything will remain energized. As long as all required safe work practices are understood and implemented, the work can be done safely and often more efficiently.

81303-12_F01.EPS

Figure 1 Transmission line work.

This module focuses on the variety of tools and equipment designed for hot stick work on energized circuits. Safe work practices related to live-line work are presented as well. Working bare-hand on transmission systems is also a common approach, and this subject is covered separately.

There are many different tasks that can potentially be performed on live circuits with hot sticks. The steps to complete common tasks, such as the movement of energized conductors, replacing cross-arms, and replacing insulators are reviewed to ensure an understanding of how the tools and safety requirements are implemented. However, keep in mind that transmission system components and procedures differ among employers, utilities, and geographic locations. Always follow the procedures provided by your employer for the area of work and the type of service being performed.

2.0.0 Live-Line Safety

It goes without saying that live-line work can be hazardous. Accidents certainly do occur on de-energized systems as well though, and it is best to practice many of the same safety habits in both cases. When transmission-level voltages are present, danger is only one error in judgment away. Although employers have a responsibility to provide a safe working environment and proper equipment, each individual worker is responsible for the safe execution of the task.

2.1.0 Job Safety Analysis

Due to the diversity of tasks and the different configurations of towers and components, a job safety analysis is always recommended. Also known as a job hazard analysis, some employers may require them to be conducted as part of a safety meeting before work begins, unless the task is routine. *OSHA Publication 3071, Job Hazard Analysis Guide*, is an excellent resource for all workers to get acquainted with the process.

2.2.0 Minimum Approach Distances

The minimum approach distance (MAD) refers to the distance that qualified personnel must maintain from an energized surface without proper PPE. These distances are documented in *OSHA Standard 1910.269(l)(2), Table R-6*. The MADs shown in *Table 1* already take into account the potential for surges in voltage. The importance of the MAD table to live-line safety cannot be overstated, so understanding the table should be one of your goals. Verifying the correct MAD any time work is being considered at a different voltage level is a good habit to develop. There is no need to rely on memory alone when planning live-line work. In fact, OSHA requires that a nonconductive plate be installed on the boom in clear sight of the bucket occupant. In addition, a nonconductive measuring device must also be within reach. The operator can use it to easily determine the distance to an energized surface.

There will be occasions when live-line workers must get closer to the task than the MAD allows. Whenever possible, this should be avoided. For obvious reasons, the potential for a shock-related injury drops dramatically when the worker is outside the MAD. Hot sticks and the many other tools available for live-line work are designed to allow tasks to be done from a safe distance. However, it is very important to note that a different set of MADs exists for hot stick work. *Table 2* shows the MAD that must be maintained when working with a hot stick. The values shown in the right column are the minimum distances that a worker's hands must be from the energized end of the hot stick.

One essential rule for working near energized surfaces in transmission systems is to protect the body from any two points of contact where current that could enter the body would also have a point of exit from the body (*Figure 2*). The exit point can be contact with ground or another phase of power. When using hot sticks, the goal is to eliminate any potential for direct contact with an energized surface. This is done by maintaining a sufficient insulation value and distance between the body and the energized surface. However, hot stick work is often done from the supporting structure of the transmission conductors. Under these circumstances, it is virtually impossible to eliminate the exit point since the tower or pole is grounded. As a result, it is essential that the MAD is maintained at all times. Hot stick workers must think ahead about their movements and ensure that the MAD's invisible barrier is not broken.

Table 1 Minimum Approach Distances

Kilovolts (AC)	Phase-to-Ground	Phase-to-Phase
0.05 to 1.0	Avoid contact	Avoid contact
1.1 to 15.0	2'-1" (0.64m)	2'-2" (0.66m)
15.1 to 36.0	2'-4" (0.72m)	2'-7" (0.77m)
36.1 to 46.0	2'-7" (0.77m)	2'-10" (0.85m)
46.1 to 72.5	3'-0" (0.90m)	3'-6" (1.05m)
72.6 to 121	3'-2" (0.95m)	4'-3" (1.29m)
138 to 145	3'-7" (1.09m)	4'-11" (1.50m)
161 to 169	4'-0" (1.22m)	5'-8" (1.71m)
230 to 242	5'-3" (1.59m)	7'-6" (2.27m)
345 to 362	8'-6" (2.59m)	12'-6" (3.80m)
500 to 550	11'-3" (3.42m)	18'-1" (5.50m)
765 to 800	14'-11" (4.53m)	26'-0" (7.91m)

From *OSHA 29 CFR 1910.269 Electric Power Generation, Transmission, and Distribution*

Table 2 Hot Stick MADs for Various Voltages

Voltage Range (Phase-to-Phase)	Minimum Working and Clear Hot Stick Distance
2.1 to 15kV	2 ft 0 in
15.1 to 35kV	2 ft 4 in
35.1 to 46kV	2 ft 6 in
46.1 to 72.5kV	3 ft 0 in
72.6 to 121kV	3 ft 4 in
138 to 145kV	3 ft 6 in
161 to 169kV	3 ft 8 in
230 to 242kV	5 ft 0 in
345 to 362kV	7 ft 0 in(1)
500 to 552kV	11 ft 0 in(1)
700 to 765kV	15 ft 0 in(1)

(1) For 345 to 362kV, 500 to 552kV, and 700 to 765kV, minimum clear hot stick distance may be reduced provided that such distances are not less than the shortest distance between the energized part and the grounded surface.

2.3.0 Live-Line PPE

A considerable amount of training has already been provided regarding the proper PPE for power line workers. However, PPE associated with hot stick live-line work is briefly reviewed here.

When working on energized systems, workers should wear fire-retardant (FR) Level 2 clothing at a minimum. Some specific tasks may require a higher level of protection. In such cases, a thorough job or task safety analysis should identify the need. The assigned level of protection for clothing is determined by the arc protection rating, stated in calories per square centimeter (cal/cm^2). To qualify as Level 2 protection, clothing must be rated to provide arc protection equal to or greater than 8.0 cal/cm^2.

Some common PPE and components, including Level 2 protection are listed here:

81303-12_F02.EPS

Figure 2 Electric shocks.

- FR long-sleeve shirt and long pants, or FR coveralls (*Figure 3*)
- 100-percent cotton undergarments
- Safety glasses with side shields, and/or polycarbonate goggles
- Nonconductive hard hat
- Nonconductive footwear

2.3.1 Hard Hats

Hard hats protect workers from a variety of hazards. They help protect against inclement weather and objects dropped from above. In live-line work, they also help prevent head contact with an energized surface.

As provided for in *ANSI/ISEA Standard Z89.1-2009,* Electrical Class E hard hats are tested to withstand 20kV. This is the class worn by power line workers. Note that the testing voltage is the same as that of dielectric footwear.

Always inspect your hard hat before using it. Never use a hard hat that has damaged webbing or a cracked outer shell. Always adjust your hard hat to fit properly before you wear it. Replace hard hats in accordance with company policy and the manufacturer's schedule.

81303-12_F03.EPS

Figure 3 Level 2 FR clothing.

2.3.2 Footwear

There are no absolute OSHA regulations regarding the footwear used for live-line work. However, OSHA does identify other widely accepted standards that protective footwear should meet. All footwear designed to prevent workplace injury, including footwear used for electrical work, should meet *ASTM International Standard F2413-05, Standard Specification for Performance Requirements for Protective Footwear.* A second standard, *ASTM International Standard F2412-05, Standard Test Methods for Foot Protection*, identifies the appropriate testing procedures to be used by footwear manufacturers. Shoes used for hot stick work should be designated as nonconductive, electrical hazard (EH) footwear (*Figure 4*). The same standard identifies the required qualities of conductive (CD) and static-dissipating (SD) footwear as well. However, hot stick workers should ensure that their shoes meet the ASTM standard for EH footwear.

EH footwear helps prevent an electrical circuit from being completed to ground, or maybe even to another conductor. Unlike gloves, footwear does not require scheduled testing once they are in service. Wear or damage that thins the sole reduces the effectiveness of EH footwear. Embedded metal in the sole, such as a tack or nail, reduces the effectiveness of EH footwear significantly. They must also be kept dry to remain effective.

Dielectric overshoes are occasionally needed, and must meet a separate standard. Under *ASTM International Standard F1117-03, Standard Specification for Dielectric Footwear,* dielectric overshoes

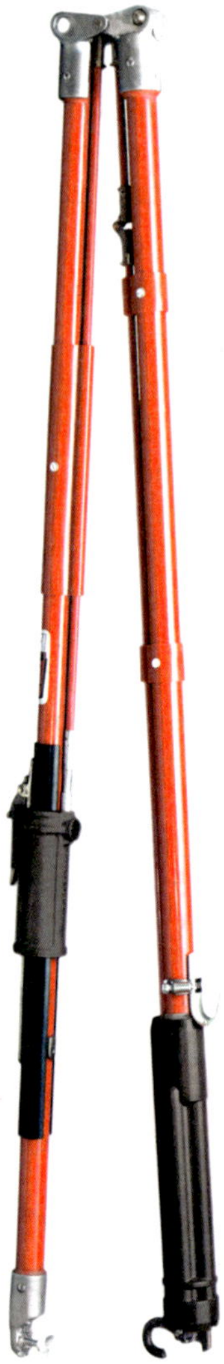

81303-12_F08.EPS

Figure 8 Hinged shotgun stick.

81303-12_F09.EPS

Figure 9 Telescoping shotgun stick.

3.2.3 Telescoping Measuring Sticks

Telescoping measuring sticks (*Figure 10*) are handy tools used to quickly determine ground clearances and measure pole heights. These sticks operate the same way as the telescoping sticks, and are available with both English and metric measuring scales. Some manufacturers offer other working sticks with measuring capability added as an option.

3.2.4 Other Dedicated Sticks

A means of using sockets to tighten hardware is often needed. A flexible insulated wrench drive (*Figure 11*) provides common square drives on each end. At the worker's end, there is a female ½" drive, while the opposite end is fitted with a ½" male drive universal joint for flexibility. A spring is added to maintain tension on the universal joint for improved control. Due to the length of the drive, the maximum torque is somewhat limited. Heavy-duty units may achieve up to 75 foot-pounds of torque, while lighter versions may be limited to 40 foot-pounds.

Hydraulic crimpers (*Figure 12*) and cutters can also be put to work at the end of a hot stick. Hydraulic hot sticks provide a rigid handle for the crimping or cutting tool, a path for the hydraulic fluid, and appropriate fittings to attach the pressure source and the tool. Most of the hydraulic tools used in live-line work require a 10,000-psig pressure source. Hydraulic tools are generally the most powerful tools.

A ratcheting wire and cable cutter (*Figure 13*) can cut ACSR conductors, as well as copper and aluminum.

81303-12_F10.EPS

Figure 10 Telescoping measuring stick.

3.2.5 Hot Stick Accessories and Attachments

A wide variety of accessories can be attached to the hot stick for specific tasks. A few common attachments are as follows:

- *Voltage detector* (*Figure 14*) – Before any live-line work begins, the voltage should be tested. The reading remains on the meter so that the worker can see it after the instrument is lowered.
- *Ammeter* (*Figure 15*) – This accessory allows workers to test the current flow in conductors, without needing to open and close the jaws of a common ammeter. As is the case with voltage, the current flow in conductors to be serviced should be checked before work begins. The open fork design allows it to be placed on the conductor easily. The reading is locked on screen until the worker has an opportunity to read the measured value.
- *Wire brush accessories* (*Figure 16*) – Before bypass jumpers or splices are applied to a conductor, the conductor surface needs to be cleaned of oxidation and corrosion that could impede current flow. Several styles are available, depending on worker preference and the typical situation encountered. The semicircle model allows for easier cleaning of the conductor on all sides, while the worker remains in a single position.
- *Conductor gauge* – It is difficult to be certain of a conductor's size from a distance. The conductor gauge (*Figure 17*) allows workers to confirm the wire sizes before preparing tools for the job. The proper selection of a number of components depends on the size of the conductor.
- *Adjustable insulator fork* (*Figure 18*) – This tool is designed to grab disk insulators. The jaws, which are constructed of a fibrous material, can be adjusted for the desired fit by turning an adjusting screw.
- *Ball socket adjusters* (*Figure 19*) – Like cotter key tools, ball socket adjusters are available in a variety of styles and shapes. The chosen style depends on the type of attaching hardware.

Figure 16

81303-12_F11.EPS

Figure 11 Flexible insulated wrench drive.

Figure 17

81303-12_F12.EPS

Figure 12 Hydraulic hot stick crimper.

4.2.4 Cross-Arm Replacement

Cross-arms are easily replaced after the conductors have been moved a safe distance away. This is best done using a boom lift and a lifting arm to control all three phases at once. This method also creates a clear work area, as opposed to using wire tongs on each phase.

Once the conductors have been moved out of the work area, the cross-arm braces are removed first. Then the cross-arm assembly itself can be removed and lowered. The replacement cross-arm should be prepared with insulators already installed at the proper distance and mounting position for the pole. This work is safer, easier, and faster if done on the ground or in the shop. Install the cross-arm with new hardware and then reinstall the cross-arm braces. The conductors can then be placed into position on the insulators and tied off.

4.3.0 Insulators

Removing and working with insulators is a common task in the power transmission environment. While some insulator strings are repaired and re-installed, others must be replaced. Porcelain insulator bells are easy to replace once the insulator is no longer physically loaded. One-piece polymer insulators cannot be repaired and are replaced when damaged. Insulators are most often repaired or replaced while the transmission system remains energized.

Most utilities and employers require that a set number of insulators in a non-polymer string be electrically sound before they can be replaced using hot sticks. This number can vary widely based on the type of structure, altitude, humidity, and a host of other factors. Workers should be certain that an insulator string targeted for service meet the requirements set forth by their employer before beginning. If it does not meet the requirements, the work must be done de-energized. Polymer insulators should be inspected for any signs of failure, most notably **carbon tracking**. The presence of corona should also be checked by listening for the tell-tale sound. When carbon tracking is present or corona can be heard, the circuit should be de-energized before the work is done.

Few pin-type insulators are found on transmission lines, since longer insulators are required due to the elevated voltage. Removing and handling insulators is not particularly complex, but the extraordinary number of pole, tower, conductor, and insulator arrangements make it challenging. Working at height adds significantly to the challenge.

Regardless of the configuration, the basic concept is consistent. The conductor must be safely captured and disconnected from the insulator. Both the insulator and conductor must be under complete control at all times. Once the conductor is disconnected and the insulator captured, the mounting hardware can be removed and the insulator can be lowered to the ground.

Rigid support insulators like the one shown in *Figure 51* are relatively simple to work with. In this case, the insulator is firmly mounted to the pole and does not need to be controlled. The conductor can be captured using wire tongs or by using the lifting arm of a boom truck. Once the insulator is free of the conductor, it can be quickly removed from its mount.

4.3.1 Suspension Insulators

Suspension insulators can be removed with several different approaches. Wire tongs from the pole or tower can be used to capture and hold the conductor temporarily while the insulator is removed. A boom truck with a lifting arm can be used to control the conductor as well. Once the insulator has been relieved of the conductor load, it can be disconnected from the suspension clamp.

For larger and heavier insulator assemblies and conductor line loads, a strain carrier can be assembled to handle the load. If a strain carrier is used, it may be connected to the same suspension

81303-12_F51.EPS

Figure 51 Rigid support insulator.

clamp or yoke that connects the insulator(s) to the conductor. *Figure 52* shows a typical yoke used with multiple suspension insulators. The front yoke of the strain carrier can be connected to the same yoke using the slotted holes. The insulators can then be disconnected from the yoke.

Although some of the work is being conducted bare-hand in the photo, *Figure 53* provides an example of a strain carrier installed on an insulator. An insulator cradle is also being used. Note that the strain carrier yokes have been connected to the two permanent yokes. The strain carrier is temporarily replacing the insulator to handle the conductor load. Since the insulator is large and cumbersome, an insulator cradle is being used to maintain control. Since suspension insulators hang vertically, J-hook assemblies to capture the insulator are of little use.

4.3.2 Dead-End and Strain Insulators

Dead-end insulators are mounted horizontally. Strain insulators are generally close to the horizontal plane as well. These configurations allow J-hook cradles to be used to capture and control the insulator. Insulator cradles can also be used.

Strain carriers can be used in these applications when necessary. If repair or replacement of the insulator strain clamp (*Figure 54*) is necessary, then a pulling grip, a strap hoist, and a link stick may be used to secure the conductor and hold it while the insulator is removed. In *Figure 55*, this approach is being used to remove two dead-end insulators at the same time. Two link sticks are connected to each other, and their connection provides a point of connection for the boom truck. Pulling grips are installed on each conductor, and a single strap hoist is being used to apply tension. After the insulators are free from the conductors, the boom truck pulls the conductors out of the work area (*Figure 56*). Note that the energized conductor loop has been captured from the ground to keep it from swinging. Once the load has been removed from the insulator, the pin can be removed from the clevis and the insulator removed.

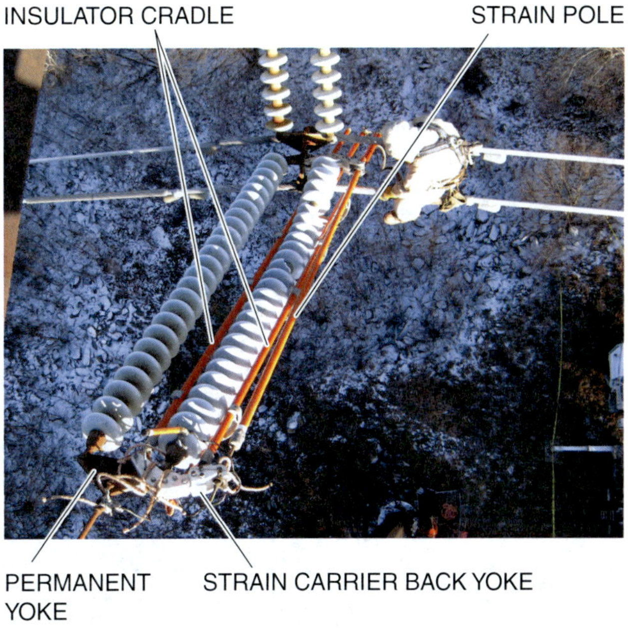

Figure 53 Strain carrier in place.

Figure 52 Suspension insulator yoke.

4.4.0 Conductor Spacers and Vibration Dampers

Conductor spacers and vibration dampers are common accessories added to transmission lines. Both devices can be installed using hot sticks.

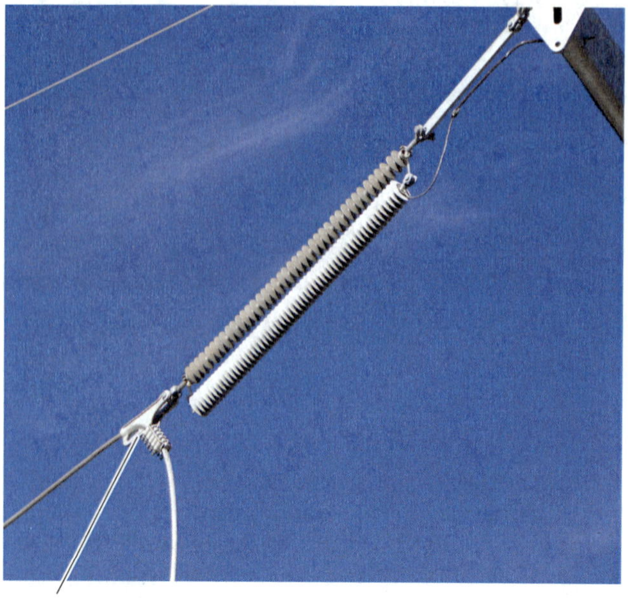

Figure 54 Conductor strain clamp.

4.4.1 Line Spacers

The line spacer shown in *Figure 57* is easily installed using shotgun sticks. The spacer assembly slides apart to allow it to be positioned on the lines, and then slides back together to capture the lines. Only one bolt is used to tighten the assembly in place. This particular device uses a break-away bolt. The top portion of the head breaks off when the proper torque is reached.

Some types of spacers can be installed with a hot stick. Always follow the manufacturer's instructors for the specific product being installed.

4.4.2 Vibration Dampers

Although there are a number of vibration damper designs on the market, those based on the original Stockbridge design (*Figure 58*) continue to be the most popular. Because of their shape, they are often referred to as dog bone dampers. A more technical name that can be used is tuned mass damper.

This style of damper is primarily designed to help prevent **aeolian vibration**, which occurs in the vertical plane. This phenomenon is identified as the primary cause of strand breakage and failure in transmission conductors. The strands at conductor terminals tend to be the most vulnerable to the effects of aeolian vibration. Although

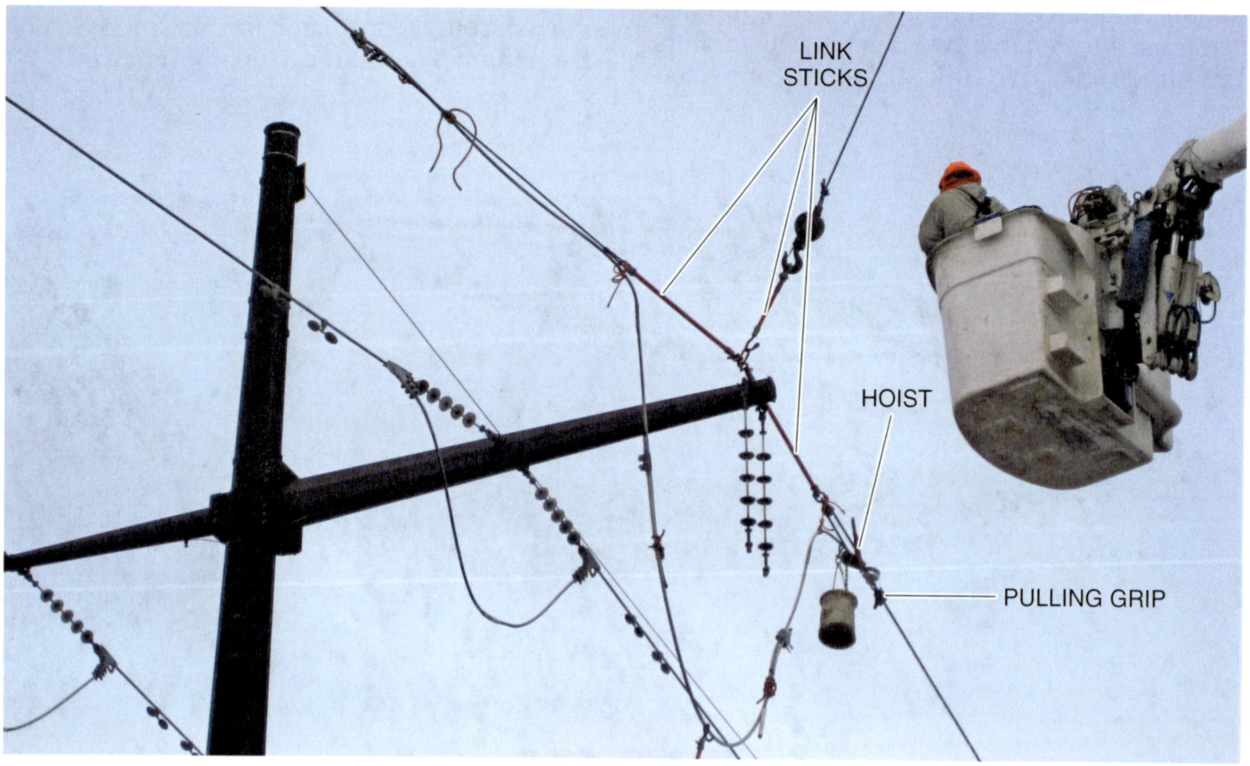

Figure 55 Pulling grip, strap hoist, and link sticks.

the basic style is consistent, some models have unequal weights at each end, slotted weights to allow for greater up/down movement, or a means of weight adjustment.

The installation of vibration dampers is quite simple with a hot stick. Workers need only position the damper at the proper line location and tighten the single bolt. They are available with break-away bolts to ensure the proper torque has been reached. The line position of the dampers must be carefully considered to create the desired effect.

81303-12_F56.EPS

Figure 56 Conductors pulled away from the work area.

81303-12_F58.EPS

Figure 58 Vibration dampers.

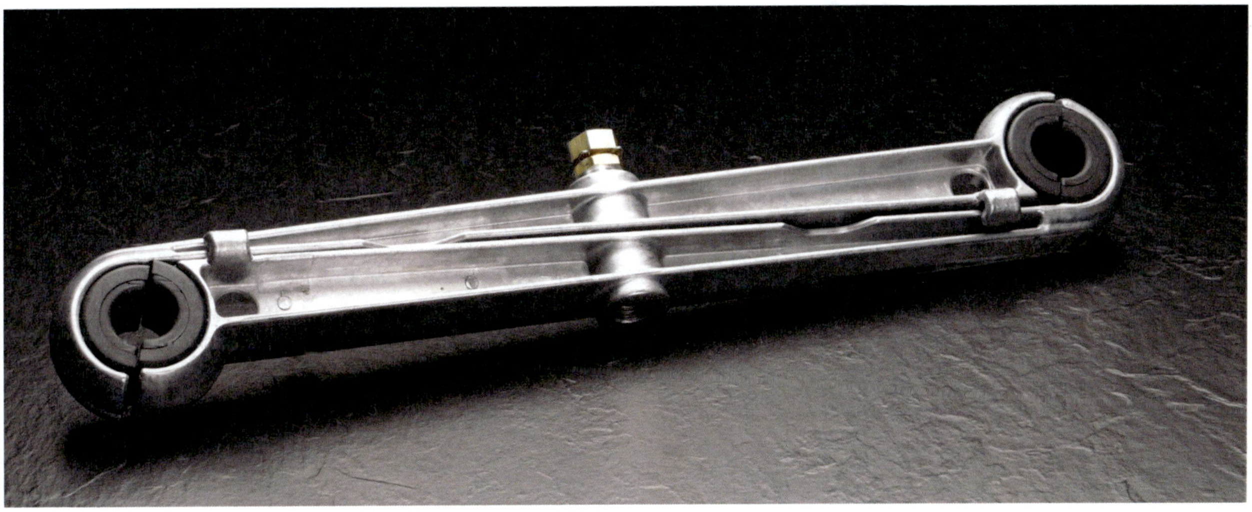

81303-12_F57.EPS

Figure 57 Line spacer.

SUMMARY

Working on energized transmission systems is a necessity today, since outages represent a hardship for both the consumer and the utility. The loss of a transmission circuit has the potential to affect large numbers of consumers. However, there is no reason why such work cannot be done in a safe and effective manner.

Live-line tools form one line of defense against worker contact with energized surfaces. A great deal of caution and attention should be given to the care and maintenance of these valuable tools. A line worker's PPE represents another line of defense. In the end however, the most effective defense every worker has is proper training and execution of the safety guidelines provided by OSHA and their employer. Careful thought and attention to every detail of live-line tasks ensure that live-line tools and PPE are in the right place at the right time, protecting workers from injury.

An array of tasks can be done using hot sticks and related tools while lines remain energized. Beyond the safe work habits and equipment required, almost every task at every pole or tower is unique and represents a different challenge. One of the keys to hot stick work is practice. Developing the skill to complete intricate work using a long stick requires dedication and patience. Teamwork is another important factor in success working with hot sticks. Many tasks require that two or more workers with hot sticks function as a team to accomplish a task.

Although a few methods of moving conductors and handling insulators have been presented here, your employer may prefer a completely different approach. These differences will not be a problem in the future for a flexible, well-trained worker who understands that safety is the first priority, regardless of the task at hand.

Review Questions

1. Level 2 clothing provides arc protection equal to or greater than _____.
   a. 4 cal/cm^2
   b. 6 cal/cm^2
   c. 8 cal/cm^2
   d. 12 cal/cm^2

2. Class E hard hats worn by power line workers are tested to withstand _____.
   a. 2,200V
   b. 5kV
   c. 10kV
   d. 20kV

3. How often does nonconductive (EH) footwear require retesting?
   a. Once each week
   b. Once each month
   c. Annually
   d. Never

4. Hot sticks made of wood must meet a higher testing standard than those constructed of FRP.
   a. True
   b. False

5. A wire tong swivel, mounted near the tip of the tong, is recommended for _____.
   a. bare conductors
   b. small conductors
   c. insulated conductors
   d. very large conductors

6. Using the equation: Holding tong force = conductor weight × (distance from the holding pole saddle to the conductor/distance between the two saddles), what is the force applied to the tong if the conductor weight is 380 pounds, the distance from the holding pole saddle to the conductor is 9 feet, and the two saddles are 6 feet apart?
   a. 63 pounds
   b. 114 pounds
   c. 253 pounds
   d. 570 pounds

7. Strain carriers can be adjusted in length after assembly using the jack screws and _____.
   a. trunnions
   b. strain poles
   c. a lever pump
   d. a hydraulic pressure source

8. Polymer insulators must be replaced when they become damaged, as they cannot be repaired.
   a. True
   b. False

9. To ensure the proper installation of a line spacer or vibration damper, some are equipped with _____.
   a. armor rods
   b. Velcro straps
   c. bolts with a break-away head
   d. nuts that can accept safety wire

10. The primary cause of strand breakage and failure in transmission conductors is _____.
    a. improperly installed insulators
    b. conductors that have become too small
    c. aeolian vibration
    d. carbon tracking

NCCER CURRICULA — USER UPDATE

NCCER makes every effort to keep its textbooks up-to-date and free of technical errors. We appreciate your help in this process. If you find an error, a typographical mistake, or an inaccuracy in NCCER's curricula, please fill out this form (or a photocopy), or complete the on-line form at **www.nccer.org/olf**. Be sure to include the exact module ID number, page number, a detailed description, and your recommended correction. Your input will be brought to the attention of the Authoring Team. Thank you for your assistance.

Instructors – If you have an idea for improving this textbook, or have found that additional materials were necessary to teach this module effectively, please let us know so that we may present your suggestions to the Authoring Team.

NCCER Product Development and Revision

13614 Progress Blvd., Alachua, FL 32615

Email: curriculum@nccer.org

Online: www.nccer.org/olf

❑ Trainee Guide ❑ AIG ❑ Exam ❑ PowerPoints Other ______________________________

Craft / Level: Copyright Date:

Module ID Number / Title:

Section Number(s):

Description:

Recommended Correction:

Your Name:

Address:

Email: Phone:

38302-11

Lift Planning

Module Four

Trainees with successful module completions may be eligible for credentialing through NCCER's National Registry. To learn more, go to **www.nccer.org** or contact us at **1.888.622.3720**. Our website has information on the latest product releases and training, as well as online versions of our *Cornerstone* newsletter and Pearson's product catalog.

Your feedback is welcome. You may email your comments to **curriculum@nccer.org**, send general comments and inquiries to **info@nccer.org**, or fill in the User Update form at the back of this module.

38302-11
Lift Planning

Objectives

When you have completed this module, you will be able to do the following:

1. Reference available material that will assist in a safe lifting operation.
2. Describe the importance of following and adhering to a lift plan.
3. Define the terms on a load/capacity chart that indicate boom angle, load radius, and boom length.
4. Calculate crane capacity using a load/capacity chart.
5. Identify the differences between on-rubber and on-outrigger charts.
6. Provide the necessary information requested on a lift plan.
7. Calculate additions and deductions involved in lifting operations.
8. Identify existing operations that need special approval.
9. Identify engineering considerations in a lift plan.
10. Identify the various types of lift plans and their differences.
11. Identify the importance of lift plan implementation.

Performance Tasks

Under the supervision of your instructor, you should be able to do the following:

1. Fill out portions of a critical lift plan as given/directed by the instructor.
2. Perform single-crane lifting calculations.
3. Perform multiple-crane lifting calculations.
4. Identify boom angle, boom length, and load radius on a load/capacity chart.
5. Identify the requirements of the on-rubber load/capacity chart.
6. Identify the requirements of the on-outrigger load/capacity chart.

Trade Terms

Boom point elevation
Engineering
Equalizer
Interpolation
Lift plan

Industry Recognized Credentials

If you're training through an NCCER-accredited sponsor you may be eligible for credentials from NCCER's Registry. The ID number for this module is 38302-11. Note that this module may have been used in other NCCER curricula and may apply to other level completions. Contact NCCER's Registry at 888.622.3720 or go to **nccer.org** for more information.

Contents

Topics to be presented in this module include:

Figures and Tables

Figures and Tables (continued)

1.0.0 Introduction

Before conducting a lift, most construction sites or companies require that a **lift plan** be completed and signed by competent personnel. Lift plans are mandatory for steel erection and multiple-crane lifts. A lift plan contains a variety of information relative to the crane, including weights, load information, and rigging requirements. It also lists any special instructions or restrictions for the lift. It is important to remember that any deviation from the original lift plan, no matter how small, requires review of the lift plan, and possibly a new lift plan.

2.0.0 Lift Plan

Critical lifts always require a lift plan. As a matter of policy, all operators should require or prepare a plan for all lifts. A typical lift plan for either a critical or ordinary (minor) lift breaks down the following elements and information (see *Appendix A*):

- Weight
  - Equipment condition
  - Weight empty
  - Weight of overhead ball
  - Weight of block
  - Weight of lifting bar
  - Weight of slings and shackles
  - Weight of jib (if erected or stowed and not used)
  - Weight of overhaul ball on jib
  - Weight of wire rope
  - Allowance for unaccounted material in equipment
- Jib
  - Jib to be used
  - Length of jib
  - Angle of jib
  - Rated capacity of jib
- Crane placement
  - Smooth solid foundation within one percent of grade
  - Electrical hazards in area
  - Obstacles or obstructions to lift or swing
  - Swing direction and degree
  - Area of operation (over-the-front, over-the-rear, over-the-side)
- Rope
  - Number of parts of rope
  - Size of rope
  - Reeving capacity
  - Sizing and types of slings
  - Sling selection
  - Shackle selection
  - Sizing and types
- Crane
  - Type of crane
  - Crane capacity
  - Lifting arrangement
- Other
  - Special instructions or restrictions for crane, rigging, or lift
  - Diagrams for crane, load placement, and rigging configuration
  - Signature of job supervisor, lift plan checker, and operator
  - Derigging plan

2.1.0 Reference Information

The personnel involved in planning the lift should consult reference information that assists in a safe lifting operation. This includes crane specifications, conditions in which the lift is taking place (weather, lift area space), and all applicable compliance and consensus standards. Standards covering mobile cranes include the following:

- *OSHA 29 CFR 1910.180*
- *OSHA 29 CFR 1926.550*
- *ASME B30.5*

3.0.0 Calculations

There are numerous calculations involved in lift planning. The personnel involved in planning the lift must take certain considerations into account:

- Available lifting-point height above the lifting location
- Load pick-up height above or below the lifting location
- Load set-down height above or below the lifting location
- Maximum wire rope weight to be experienced during the lift
- Weight of load to be lifted, not including rigging
- Weight of crane components below the lifting point (deductions)
  - Length of line from point to load block
  - Any parts of line that are considered part of the load
  - Load block height
  - Rigging on the block, including lifting beams, slings, and shackles
- Effective weight of crane components past the lifting point
  - Jib
  - Line to jib top
  - Length of line from point to load block

- Weight of any additional crane components and/or boom components

3.1.0 Single-Crane Lifting

The following information is for instructional purposes only. The load/capacity charts presented here are derived from manufacturer sales information that may not be complete or machine-specific. They are intended to give you practice in lift calculations. On a job site, always follow the manufacturer's load charts and recommendations.

To illustrate lift plan calculations, consider this example. What is the net capacity of the Manitowoc 4100W-2 in the following configuration?

- 22C boom, 190-ft main boom, no offset
- 55 ft of load radius
- Two-sheave boom point attached
- 15-ton hook block 20 ft below boom point
- 100-ton block reeved with 4 parts of line
- Crawlers extended
- Rigging of 365 lbs
- Lifting from the main boom

Refer to the Manitowoc 4100W-2 load/capacity charts shown in *Figures 1* through *3*.

The gross capacity of the crane at 190 ft of boom length and 55 ft of load radius is 88,000 lbs. Deductions from this weight include:

- Two-sheave boom point = 1,500 lbs
- 15-ton hook block or hook and weight ball = 865 lbs
- 100-ton block = 2,065 lbs
- Rigging = 365 lbs
- Rope = 1,827 lbs

The total weight of the deductions is 6,622 lbs. The net capacity of the crane in this configuration is the gross capacity (88,000 lbs) minus the deductions (6,622 lbs), which is 81,378 lbs.

3.2.0 Multiple-Crane Lifting

Multiple-crane lifts are always considered critical lifts, requiring detailed plans and procedures. The effects of the relative motion between cranes may induce additional weight on the cranes, the load, and the lifting gear. Multiple-crane lifting must be used when the physical dimensions, characteristics, mass, or required movement of the load prevent the operation from being carried out by a single crane. Procedures must be in the written form of a lift plan and signed by qualified personnel. These factors should be considered when planning multiple-crane lifts:

- The lift must be planned and carried out under qualified supervision.
- Ground conditions must be stable and level.
- Consideration must be given to the environment and the location of the lift.
- All cranes must rest on a solid, firm, level surface.
- All cranes must be level as per manufacturer's recommendations.
- The load weight must be known.
- The longest load radius for each crane of the operation must be measured.
- The maximum allowable boom length operating radius and minimum boom angle must be known.
- The rated capacity of each crane must be known.
- Because loads transfer rapidly, no crane should be loaded to more than 75 percent of its net capacity.
- Exactly how much of the load to be carried by each crane must be known. The rigging must be arranged to divide the load as planned.
- The line, swing, and boom speeds of the cranes must be matched.
- Swing and booming methods should be kept to a minimum.
- Whenever possible, the cranes should not travel with the load.
- A lift plan must be developed.
- Signal persons, riggers, and operators must know exactly what to do before the lift begins.
- If possible, a dry run should be conducted without a load.
- Communications by radio are preferred when hand signals are not effective.
- One person is required to direct and control the operation.
- Hoist lines must be kept vertical at all times. When the lines are not vertical, the cranes are transferring load to each other, and one crane may sideload the other.

WARNING!

Multiple-crane lifts are particularly hazardous. Because the movements of multiple cranes must be closely coordinated, all crane movements must be carefully planned and slowly executed to prevent lift failures. All personnel involved must be thoroughly familiar with the lift plan and any applicable emergency procedures.

Multiple cranes are often used to lift very heavy loads. *Figure 4* shows a three-crane lift with two of the cranes using an **equalizer** beam with a special load attachment. The equalizer beam and attachment split the load between the two cranes. Equalizer beams are covered later in this module.

MANITOWOC 4100 W-2

LIFTCRANE CAPACITIES

BOOM NO. 22C WITH OPEN THROAT TOP
146,400 LB. CRANE COUNTERWEIGHT
60,000 LB. CARBODY COUNTERWEIGHT
26'6" CRAWLERS EXTENDED

WARNING: This chart will apply only when two 12,000 lb. side ctwts. and two 30,000 lb. carbody ctwts. bear MEC registered Serial Numbers.

LIFTING CAPACITIES: Capacities for various boom lengths and operating radii may be based on percent of tipping, strength of structural components, operating speeds and other factors.

Capacities are for freely suspended loads and do not exceed 75% of a static tipping load. Capacities based on structural competence are shown by shaded areas.

Capacities are shown in pounds. Deduct 1200 pounds from capacities listed when single sheave upper boom point is attached and 1500 pounds when two sheave upper boom point is attached. To comply with B30.5 requirements, upper boompoint cannot be used on the 260 ft. boom. Weight of jib, (see chart A), all load blocks, hooks, weight ball, slings, hoist lines beneath boom and jib point sheaves, etc., is considered part of the main boom load. Boom is not to be lowered beyond radii where combined weights are greater than rated capacity. Where no capacity is shown, operation is not intended or approved.

OPERATING CONDITIONS: Machine to operate in a level position on a firm surface with crawlers fully extended and gantry in working position and be rigged in accordance with and under conditions referred to in rigging drawing No. 190693 and load line specification chard No. 6592-A.

Crane operator judgement must be used to allow for dynamic load effects of swinging, hoisting or lowering, travel, as well as adverse operating conditions & physical machine depreciation.

OPERATOR RADIUS: Operating is the horizontal distance from the axis of rotation to the center of vertical hoist line or load block with the load freely suspended. Add 14" to boom point radius for radius of sheave when using single part hoist line.

Boom angle is the angle between horizontal and centerline of boom butt and inserts and is an indication of operating radius. In all cases, operating radius shall govern capacity.

BOOM POINT ELEVATION: Boom point elevation, in feet, is the vertical distance from ground level to centerline of boom point shaft.

MACHINE EQUIPMENT: Machine equipped with 26'6" extendible crawlers, 48" treads, 17' retractable gantry, 12 part boom hoist reeving, four 1 3/8" boom pendants, 1st ctwt. 41,900 lbs., 2nd ctwt. 41,500 lbs., 3rd ctwt. 39,000 lbs., two 12,000 lbs. side ctwt's. and two 30,000 lbs. carbody ctwt's.

LOAD AND WHIP LINE SPECIFICATIONS	
LOAD LINE:	1-1/8" - 6 x 31 Warrington-Seale, Extra Improved Plow Steel, Regular Lay, IWRC. Minimum Breaking Strength 65 Ton. (Approx.Weight Per Ft. in Lbs. 2.34)
WHIP LINE:	1-1/8" - Warrington-Seale, Improved Plow Steel, Regular Lay, IWRC. Minimum Breaking Strength 56.5 Ton. Maximum Load - 28,300 Lbs. Per Line. (Approx. Weight Per Ft. in Lbs. 2.34)

MAXIMUM BOOM AND JIB LENGTH LIFTED UNASSISTED				DEDUCT FROM CAPACITIES WHEN JIB IS ATTACHED	
OVER FRONT OF BLOCKED CRAWLERS		OVER SIDE OF EXTENDED CRAWLERS			
BOOM LENGTH	JIB NO. 123	BOOM LENGTH	JIB NO. 123	JIB LENGTH	JIB NO. 123
260'	--	260'	--	30'	3,000 lbs.
250'	--	250'	--	40'	3,600 lbs.
240'	40'	240'	40'	50'	4,200 lbs.
230'	60'	230'	60'	60'	4,900 lbs.
Load block, hook and weight ball on ground to start.					

HOIST REEVING FOR MAIN LOAD BLOCK						
No.Parts Of Line	1	2	3	4	5	6
Max Load - Lbs.	32,500	65,000	97,500	130,000	162,500	195,000
No. Parts of Line	7	8	9	10	11	12
Max. Load - Lbs.	227,500	260,000	292,500	325,000	357,500	400,000
No. Parts of Line	13					
Max. Load - Lbs.	430,000					

FOR JIB CAPACITIES, CONSULT JIB CHART.

BOOM LGTH FEET	OPER. RAD. FEET	BOOM ANG. DEG.	BOOM POINT ELEV.	CAPACITY: CRAWLERS EXTENDED
70	16.5	79.7	75.9	460,000
	17	79.3	75.8	400,000
	18	78.5	75.6	380,100
	19	77.6	75.4	363,000
	20	76.8	75.1	347,300
	22	75.1	74.6	319,600
	24	73.4	74.1	293,400
	26	71.7	73.5	266,100
	28	69.9	72.8	237,500
	30	68.2	72.0	214,300
	32	66.4	71.2	195,100
	34	64.6	70.2	178,900
	36	62.8	69.3	165,200
	38	60.9	68.2	153,300
	40	59.1	67.0	143,000
	45	54.1	63.7	122,100
	50	48.9	59.8	106,300
	55	43.2	54.9	93,900
	60	36.9	49.0	84,000
	65	29.4	41.3	75,800
	70	19.5	30.3	63,900

BOOM LGTH FEET	OPER. RAD. FEET	BOOM ANG. DEG.	BOOM POINT ELEV.	CAPACITY: CRAWLERS EXTENDED
80	17	80.6	85.9	392,800
	18	79.9	85.8	378,900
	19	79.2	85.6	361,800
	20	78.5	85.4	346,100
	22	77.0	84.9	318,400
	24	75.5	84.5	292,500
	26	74.0	83.9	265,600
	28	72.5	83.3	237,000
	30	71.0	82.7	213,800
	32	69.5	81.9	194,600
	34	68.0	81.2	178,500
	36	66.4	80.3	164,700
	38	64.8	79.4	152,800
	40	63.3	78.4	142,400
	45	59.2	75.7	121,500
	50	54.9	72.5	105,700
	55	50.4	68.6	93,300
	60	45.6	64.1	83,400
	65	40.3	58.8	75,200
	70	34.4	52.2	68,300
	75	27.4	43.9	52,500
	80	18.2	32.0	53,900

BOOM LGTH FEET	OPER. RAD. FEET	BOOM ANG. DEG.	BOOM POINT ELEV.	CAPACITY: CRAWLERS EXTENDED
90	18	81.1	95.9	355,400
	19	80.4	95.7	346,900
	20	79.8	95.6	336,900
	22	78.5	95.2	317,400
	24	77.2	94.7	291,700
	26	75.9	94.3	264,800
	28	74.5	93.7	236,600
	30	73.2	93.2	213,400
	32	71.9	92.5	194,200
	34	70.5	91.9	178,000
	36	69.2	91.1	164,200
	38	67.8	90.3	152,300
	40	66.4	89.5	142,000
	45	62.9	87.1	121,100
	50	59.3	84.4	105,200
	55	55.5	81.2	92,800
	60	51.5	77.5	82,900
	65	47.3	73.2	74,700
	70	42.8	68.2	67,800
	75	37.9	62.3	62,000
	80	32.4	55.2	57,000
	85	25.8	46.2	52,600
	90	17.1	33.5	45,900

BOOM LGTH FEET	OPER. RAD. FEET	BOOM ANG. DEG.	BOOM POINT ELEV.	CAPACITY: CRAWLERS EXTENDED
100	19	81.4	105.9	332,900
	20	80.8	105.7	327,100
	22	79.6	105.4	316,200
	24	78.5	105.0	290,800
	26	77.3	104.5	263,900
	28	76.1	104.1	236,200
	30	74.9	103.6	212,900
	32	73.7	103.0	193,700
	34	72.5	102.4	177,500
	36	71.3	101.7	163,700
	38	70.1	101.0	151,800
	40	68.9	100.3	141,400
	45	65.8	98.2	120,500
	50	62.6	95.8	104,700
	55	59.3	93.0	92,300
	60	55.9	89.8	82,300
	65	52.4	86.2	74,100
	70	48.7	82.1	67,200
	75	44.8	77.4	61,400
	80	40.5	72.0	56,400
	85	35.9	65.6	52,000
	90	30.7	58.0	48,200
	95	24.5	48.5	44,900
	100	16.3	35.0	39,300

CAUTION! CHECK AMOUNT OF COUNTERWEIGHT ON MACHINE BEFORE USE OF THIS CHART

38302-11_F01.EPS

Figure 1 Load/capacity chart, page one.

LIFTCRANE CAPACITIES – 4100W SERIES 2 BOOM NO 22C WITH OPEN-THROAT TOP, CONTINUED

BOOM LGTH FEET	OPER. RAD. FEET	BOOM ANG. DEG.	BOOM POINT ELEV.	CAPACITY: CRAWLERS EXTENDED
110	22	82.3	115.7	240,500
	24	81.3	115.4	232,100
	26	80.2	115.0	224,400
	28	79.1	114.6	217,300
	30	78.1	114.1	210,700
	32	77.0	113.6	193,300
	34	75.9	113.1	177,100
	36	74.8	112.5	163,300
	38	73.7	111.8	151,400
	40	72.6	111.2	141,000
	45	69.9	109.3	120,100
	50	67.0	107.2	104,200
	55	64.1	104.7	91,800
	60	61.2	101.9	81,800
	65	58.1	98.8	73,600
	70	54.9	95.3	66,800
	75	51.6	91.3	60,900
	80	48.1	86.8	55,900
	85	44.4	81.7	51,600
	90	40.4	75.9	47,800
	95	36.1	69.2	44,400
	100	31.1	61.1	41,400
	105	25.3	51.1	38,700
	110	17.6	37.2	33,900
120	22	83.0	125.9	228,700
	24	82.0	125.5	220,500
	26	81.0	125.2	213,000
	28	80.0	124.8	206,100
	30	79.1	124.4	199,700
	32	78.1	123.9	192,800
	34	77.1	123.4	176,600
	36	76.1	122.9	162,800
	38	75.1	122.3	150,900
	40	74.1	121.7	140,500
	45	71.6	120.0	119,500
	50	69.1	118.1	103,700
	55	66.5	115.9	91,300
	60	63.8	113.4	81,300
	65	61.1	110.6	73,100
	70	58.3	107.5	66,200
	75	55.4	104.0	60,400
	80	52.4	100.1	55,300
	85	49.2	95.8	51,000
	90	45.9	91.0	47,200
	95	42.4	85.5	43,800
	100	38.6	79.4	40,800
	105	34.4	72.2	38,100
	110	29.7	63.7	35,700
	115	24.1	53.2	33,500
130	24	82.6	135.7	212,700
	26	81.7	135.3	205,300
	28	80.8	135.0	198,600
	30	79.9	134.6	192,300
	32	79.0	134.2	186,500
	34	78.1	133.7	176,300
	36	77.2	133.2	162,500
	38	76.3	132.7	150,600
	40	75.4	132.1	140,200
	45	73.1	130.6	119,300
	50	70.8	128.8	103,400
	55	68.4	126.8	91,000
	60	66.0	124.6	81,000
	65	63.5	122.0	72,800
	70	61.0	119.2	65,900
	75	58.4	116.1	60,000
	80	55.7	112.7	55,000
	85	53.0	108.9	50,700
	90	50.1	104.8	46,800
	95	47.1	100.1	43,500
	100	43.9	95.0	40,500
	105	40.6	89.2	37,800
	110	36.9	82.6	35,300
	115	33.0	75.1	33,100
	120	28.5	66.1	31,100
	125	23.1	55.1	29,300
140	26	82.3	145.5	198,400
	28	81.5	145.1	191,800
	30	80.7	144.8	185,600
	32	79.8	144.4	179,900
	34	79.0	144.0	174,700
	36	78.1	143.5	162,000
	38	77.3	143.0	150,100
	40	76.5	142.5	139,700
	45	74.3	141.1	118,700
	50	72.2	139.4	102,800
	55	70.0	137.6	90,400
	60	67.8	135.5	80,400
	65	65.6	133.2	72,200
	70	63.3	130.7	65,300
	75	60.9	127.9	59,500
	80	58.5	124.8	54,400
	85	56.1	121.4	50,100
	90	53.5	117.7	46,200
	95	50.9	113.7	42,900
	100	48.1	109.2	39,900
140	105	45.3	104.2	37,200
	110	42.2	98.8	34,700
	115	39.0	92.7	32,500
	120	35.5	85.7	30,500
	125	31.7	77.8	28,700
	130	27.4	68.5	27,000
	135	22.2	57.0	25,500
150	28	82.1	155.3	185,600
	30	81.3	154.9	179,600
	32	80.5	154.6	174,000
	34	79.7	154.2	168,800
	36	79.0	153.8	161,600
	38	78.2	153.3	149,600
	40	77.4	152.8	139,200
	45	75.4	151.5	118,200
	50	73.4	150.0	102,400
	55	71.4	148.3	89,900
	60	69.4	146.4	79,900
	65	67.3	144.3	71,700
	70	65.2	141.9	64,800
	75	63.0	139.4	59,000
	80	60.9	136.6	53,900
	85	58.6	133.5	49,600
	90	56.3	130.1	45,700
	95	54.0	126.5	42,400
	100	51.5	122.5	39,400
	105	49.0	118.2	36,700
	110	46.4	113.4	34,200
	115	43.6	108.2	32,000
	120	40.7	102.4	30,000
	125	37.6	96.0	28,200
	130	34.2	88.8	26,500
	135	30.5	80.5	25,000
	140	26.4	70.8	23,500
	145	21.4	58.8	22,200
160	28	82.6	165.4	179,800
	30	81.8	165.1	173,900
	32	81.1	164.7	168,400
	34	80.4	164.4	163,300
	36	79.7	164.0	158,600
	38	78.9	163.6	149,100
	40	78.2	163.1	138,700
	45	76.3	161.9	117,700
	50	74.5	160.5	101,800
	55	72.6	158.9	89,300
	60	70.7	157.1	79,300
	65	68.8	155.1	71,100
	70	66.8	153.0	64,200
	75	64.9	150.6	58,400
	80	62.9	148.0	53,300
	85	60.8	145.2	49,000
	90	58.7	142.2	45,100
	95	56.6	138.8	41,800
	100	54.4	135.2	38,700
	105	52.1	131.4	36,000
	110	49.7	127.1	33,600
	115	47.3	122.5	31,400
	120	44.8	117.5	29,400
	125	42.1	112.0	27,600
	130	39.3	105.9	25,900
	135	36.3	99.2	24,300
	140	33.1	91.7	22,900
	145	29.5	83.0	21,600
	150	25.5	72.9	20,400
	155	20.7	60.5	19,200
170	30	82.3	175.2	169,000
	32	81.6	174.9	163,600
	34	81.0	174.5	158,700
	36	80.3	174.2	154,000
	38	79.6	173.8	148,800
	40	78.9	173.3	138,400
	45	77.2	172.2	117,400
	50	75.4	170.9	101,500
	55	73.7	169.4	89,100
	60	71.9	167.7	79,000
	65	70.1	165.9	70,800
	70	68.3	163.9	63,900
	75	66.4	161.7	58,100
	80	64.6	159.3	53,000
	85	62.7	156.7	48,700
	90	60.8	153.9	44,900
	95	58.8	150.8	41,500
	100	56.8	147.5	38,500
	105	54.7	144.0	35,800
	110	52.6	140.2	33,300
	115	50.4	136.0	31,100
	120	48.1	131.5	29,100
	125	45.8	126.7	27,300
	130	43.3	121.4	25,600
	135	40.8	115.7	24,100
	140	38.1	109.3	22,600
	145	35.2	102.3	21,300
	150	32.0	94.5	20,100
	155	28.6	85.5	18,900
	160	24.7	75.1	17,800
	165	20.0	62.2	16,400
180	32	82.1	185.0	158,700
	34	81.5	184.7	154,000
	36	80.8	184.3	149,500
	38	80.2	184.0	145,200
	40	79.5	183.6	137,900
	45	77.9	182.5	116,900
	50	76.3	181.2	101,000
	55	74.6	179.8	88,500
	60	72.9	178.3	78,500
	65	71.3	176.6	70,300
	70	69.6	174.7	63,400
	75	67.8	172.6	57,500
	80	66.1	170.4	52,500
	85	64.3	168.0	48,100
	90	62.5	165.4	44,300
	95	60.7	162.5	40,900
	100	58.9	159.5	37,900
	105	57.0	156.2	35,200
	110	55.0	152.7	32,700
	115	53.0	149.0	30,500
	120	51.0	144.9	28,500
	125	48.9	140.5	26,700
	130	46.7	135.8	25,000
	135	44.4	130.7	23,500
	140	42.0	125.2	22,000
	145	39.6	119.2	20,700
	150	36.9	112.6	19,500
	155	34.1	105.3	18,300
	160	31.1	97.2	17,200
	165	27.7	87.9	16,200
	170	23.9	77.1	15,300
	175	19.4	63.8	13,500
190	32	82.5	195.1	153,100
	34	81.9	194.8	148,800
	36	81.3	194.5	144,800
	38	80.7	194.1	140,800
	40	80.1	193.8	137,000
	45	78.5	192.7	116,400
	50	77.0	191.5	100,500
	55	75.4	190.2	88,000
	60	73.9	188.8	78,000
	65	72.3	187.1	69,700
	70	70.7	185.4	62,800
	75	69.1	183.4	57,000
	80	67.4	181.4	52,000
	85	65.8	179.1	47,600
	90	64.1	176.6	43,800
	95	62.4	174.0	40,400
	100	60.7	171.2	37,400
	105	58.9	168.2	34,700
	110	57.1	164.9	32,200
	115	55.3	161.4	30,000
	120	53.4	157.7	28,000
	125	51.5	153.7	26,200
	130	49.5	149.5	24,500
	135	47.5	144.9	22,900
	140	45.3	140.0	21,500
	145	43.1	134.6	20,200
	150	40.8	128.9	18,900
	155	38.4	122.6	17,800
	160	35.9	115.8	16,700
	165	33.2	108.3	15,700
	170	30.2	99.8	14,800
	175	27.0	90.3	13,900
	180	23.3	79.1	13,100
	185	18.9	65.4	11,000
200	34	82.3	204.9	143,100
	36	81.7	204.6	139,500
	38	81.2	204.3	136,000
	40	80.6	203.9	132,500
	45	79.1	203.0	115,900
	50	77.7	201.8	100,000
	55	76.2	200.6	87,500
	60	74.7	199.2	77,400
	65	73.2	197.7	69,200
	70	71.7	196.0	62,300
	75	70.2	194.2	56,400
	80	68.6	192.2	51,400
	85	67.1	190.1	47,000
	90	65.5	187.8	43,200
	95	63.9	185.3	39,800
	100	62.3	182.7	36,800
	105	60.6	179.8	34,100
	110	59.0	176.8	31,600
	115	57.3	173.6	29,400
	120	55.5	170.2	27,400
	125	53.0	166.5	25,600
	130	51.9	162.6	23,900
	135	50.1	158.4	22,300
	140	48.1	153.9	20,900
	145	46.2	149.1	19,600
	150	44.1	144.0	18,300
	155	42.0	138.4	17,200
	160	39.7	132.5	16,100
	165	37.4	126.0	15,100
	170	34.9	118.9	14,200
200	175	32.3	111.1	13,300
	180	29.4	102.4	12,400
	185	26.2	92.5	11,700
	190	22.6	81.0	10,900
	195	18.4	67.0	8,500
210	36	82.1	214.7	134,200
	38	81.6	214.4	131,000
	40	81.0	214.1	127,900
	45	79.6	213.2	115,500
	50	78.3	212.1	99,600
	55	76.9	210.9	87,100
	60	75.4	209.6	77,100
	65	74.0	208.1	68,800
	70	72.6	206.6	61,900
	75	71.2	204.8	56,100
	80	69.7	203.0	51,000
	85	68.2	201.0	46,700
	90	66.7	198.8	42,800
	95	65.2	196.5	39,400
	100	63.7	194.0	36,400
	105	62.2	191.3	33,700
	110	60.6	188.5	31,300
	115	59.0	185.5	29,100
	120	57.4	182.3	27,100
	125	55.7	178.9	25,200
	130	54.1	175.2	23,500
	135	52.3	171.4	22,000
	140	50.6	167.3	20,600
	145	48.8	162.9	19,200
	150	46.9	158.2	18,000
	155	45.0	153.2	16,800
	160	43.0	147.9	15,800
	165	40.9	142.1	14,800
	170	38.7	135.9	13,800
	175	36.5	129.2	12,900
	180	34.0	121.9	12,100
	185	31.5	113.9	11,300
	190	28.7	104.9	10,600
	195	25.6	94.7	9,900
	200	22.1	82.9	8,800
	205	17.9	68.5	6,500
220	38	82.0	224.6	125,700
	40	81.4	224.2	122,800
	45	80.1	223.3	115,000
	50	78.8	222.3	99,100
	55	77.5	221.2	86,600
	60	76.1	220.0	76,600
	65	74.8	218.6	68,300
	70	73.4	217.1	61,400
	75	72.0	215.4	55,500
	80	70.7	213.7	50,500
	85	69.3	211.8	46,100
	90	67.9	209.7	42,300
	95	66.4	207.5	38,900
	100	65.0	205.2	35,800
	105	63.6	202.7	33,100
	110	62.1	200.0	30,700
	115	60.6	197.2	28,500
	120	59.1	194.2	26,500
	125	57.5	191.0	24,600
	130	55.9	187.6	23,000
	135	54.3	184.0	21,400
	140	52.7	180.2	20,000
	145	51.0	176.1	18,600
	150	49.3	171.8	17,400
	155	47.6	167.3	16,300
	160	45.7	162.4	15,200
	165	43.9	157.2	14,200
	170	41.9	151.7	13,200
	175	39.9	145.7	12,300
	180	37.8	139.3	11,500
	185	35.6	132.4	10,700
	190	33.2	124.8	10,000
	195	30.7	116.6	9,300
	200	28.0	107.3	8,600
	205	25.0	96.9	8,000
	210	21.5	84.8	6,500

CAPACITIES CONTINUED ON NEXT PAGE

These load charts are intended for instructional purposes only. They were derived from manufacturer sales information which may not be complete or machine specific. Not responsible for typographical errors.

38302-11_F02.EPS

Figure 2 Load/capacity chart, page two.

WEIGHTS 4100W Cont'd

DESCRIPTION	APPROX. WEIGHT (IN LBS.)
JIB NO. 123	
Jib Top - 15' (w/Jib point)	695
Jib Butt - 15'	690
Jib Insert - 10'	340
Basic Pendant - 33' 3-3/4" (2 Req'd)	115 each
Pendant - 10' (2 Per insert)	65 each
Jib Backstay Pendant	155 each
Jib Strut - 12'-6"	365
COMPONENTS	
Hook Rollers (6) - w/shafts	1,020
Light Plant - 6.5KW - w/mounting platform	1,390
Catwalk - left and right side w/rails	1,320
Lagging - 27-5/8" dia. plain	1,410
Boom Hoist Rope - 12 Part - 760' of 7/8" - 6 x 26	1,080
Wire Rope Guide Assembly - Lower	325
Wire Rope Guide Assembly - Upper	510
Rope Guide Roller Assembly	55 each
2-Part Gantry w/Telescopic Backhitch	7,805
Equalizer	2,000
Hoist Line - 1-1/8" - 6 x 31	2.34 lbs./ft.
Whip Line - 1-1/8" - 6 x 31	2.34 lbs./ft.
15-Ton Hook and Weight Ball	865
100-Ton Hook Block Assembly	2,065
200-Ton Hook Block Assembly	4,900
230-Ton Hook Block Assembly	5,375
Boom Stop - Telescopic Air Cushioned	675
Dragline Fairlead - Revolving	1,910
Dragline Fairlead - Hinged	9,330

NOTE: The above weights may fluctuate up or down 5% due to manufacturing tolerances.

38302-11_F03.EPS

Figure 3 Load/capacity chart, page three.

Whenever two or more cranes are used to lift a load, the location of each crane's hook block above the load has to be chosen carefully so that the load is shared between the cranes. Because each crane used will usually have a different net capacity, the method for properly locating each crane's hook block above the load (*Figure 5*) for a two-crane lift is as follows:

Step 1 Determine the lowest value of rated net capacity that Crane One will have during the whole operation. When lifting from the main boom or an extension, consider all factors, including configuration, quadrant of operation, boom length, boom angle, load radius, weight of rigging, and all capacity deductions.

38302-11_F04.EPS

Figure 4 Three-crane lift.

Step 2 Repeat Step 1 for Crane Two.

Step 3 Ensure that the sum of the lowest rated net capacities of the cranes exceeds the load weight.

Step 4 Locate the position of the load's center of gravity.

Step 5 Each crane's load blocks must be positioned relative to the load's center of gravity according to the following formula:

$$(\text{net capacity of Crane One}) \times A = (\text{net capacity of Crane Two}) \times B$$

Step 6 The crane with the greatest net capacity must always be closest to the load's center of gravity. Measure the distance of this point from the center of gravity. If you have measured distance A, then:

$$B = \frac{\text{net capacity of Crane One}}{\text{net capacity of Crane Two}} \times A$$

If you have measured distance B then:

$$A = \frac{\text{net capacity of Crane Two}}{\text{net capacity of Crane One}} \times B$$

Step 7 Determine the share of the load weight that will be carried by each crane:

$$\text{Load on Crane One} = \frac{B}{A + B} \times \text{load weight}$$

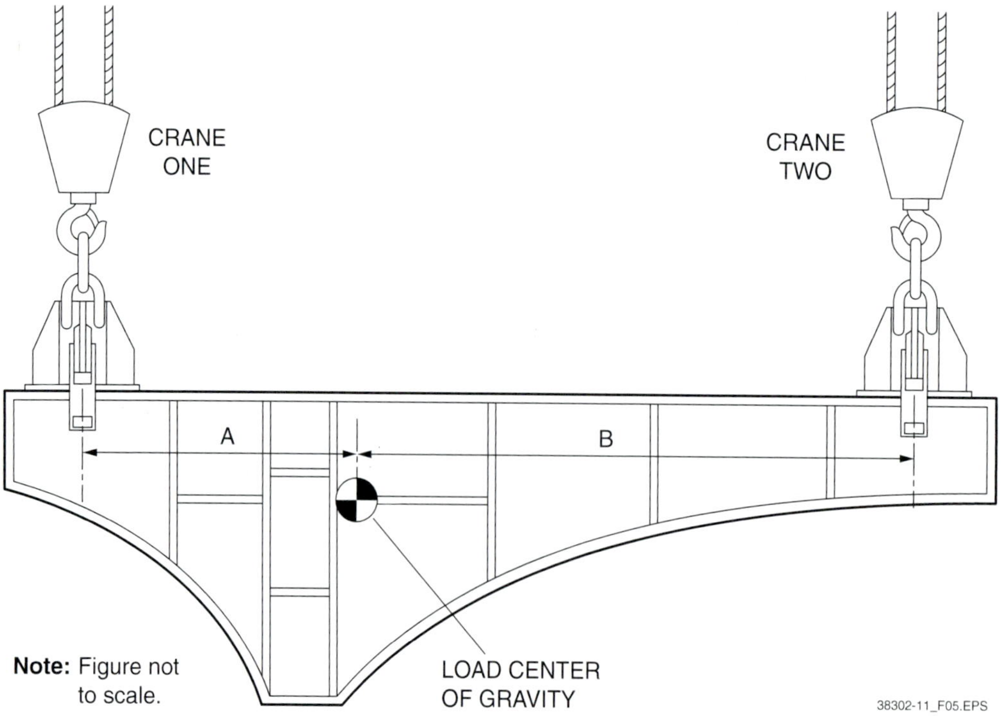

Figure 5 Multiple-crane lift.

$$\text{Load on Crane Two} = \frac{A}{A + B} \times \text{load weight}$$

Step 8 Ensure that neither crane is loaded beyond its lowest rated net capacity.

> **NOTE**
> Each lifting operation is unique. The numbers provided in this example are randomly selected for instructional purposes.

To calculate that neither crane in the example is loaded beyond its lowest rated net capacity, follow these steps:

Step 1 Lowest rated net capacity of Crane One = 156,000 lbs – 500 lbs (beam clamp) = 155,500 lbs.

Step 2 Lowest rated net capacity of Crane Two = 90,500 lbs – 500 lbs (beam clamp) = 90,000 lbs.

Step 3 Net capacity = 155,500 lbs + 90,000 lbs = 245,500 lbs.

Step 4 Locate the load's center of gravity from the ends of the load, as shown in *Figure 6*.

Step 5 To set the location of the hook blocks, use the formula:

$$(\text{Net capacity of Crane One}) \times A = (\text{net capacity of Crane Two}) \times B$$

Step 6 Choose a hypothetical length for A that will be within the length of C in *Figure 6*. Start with A = 45 ft (*Figure 7*), for example.

$$\text{Calculate } B = \frac{\text{net capacity of Crane One}}{\text{net capacity of Crane Two}} \times A$$

$$= \frac{155{,}500 \text{ lbs}}{90{,}000 \text{ lbs}} \times 45 \text{ ft} = 78 \text{ ft}$$

This configuration will not work because the hook block for B (78 ft) would be beyond the end of the right side of the load (*Figure 8*). Choose another dimension for A: 36 ft (*Figure 9*).

$$\text{Recalculate } B = \frac{\text{net capacity of Crane One}}{\text{net capacity of Crane Two}} \times A$$

$$= \frac{155{,}500 \text{ lbs}}{90{,}000 \text{ lbs}} \times 36 \text{ ft} = 62.2 \text{ ft}$$

This dimension fits the load (*Figure 10*).

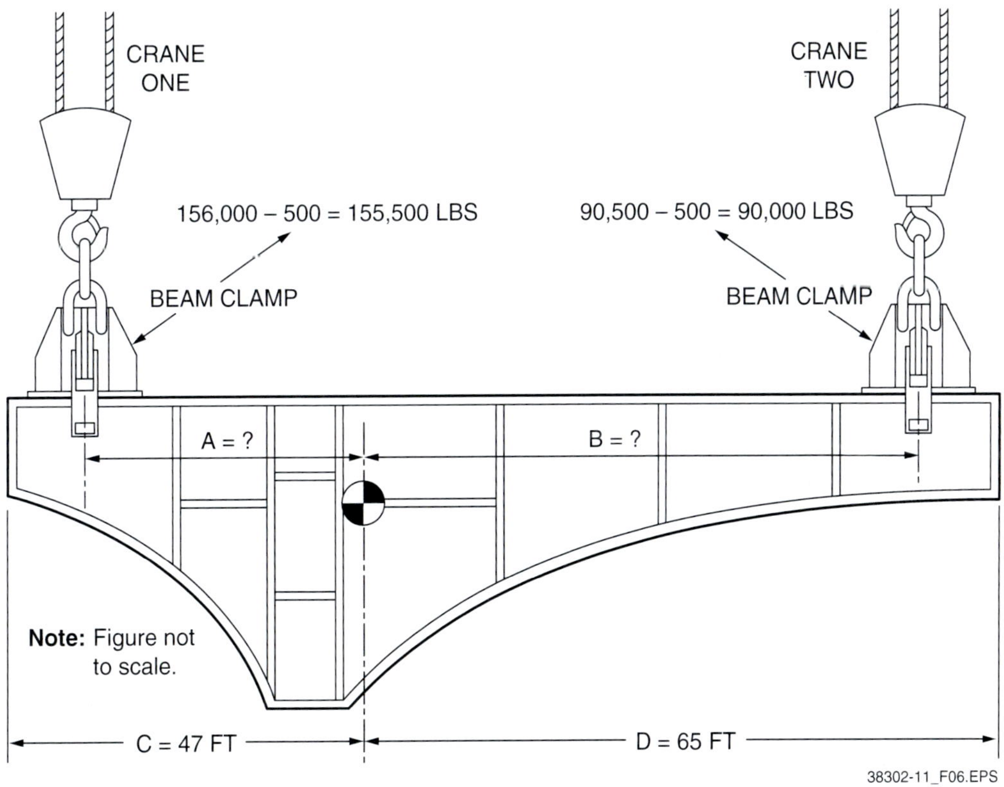

Figure 6 Multiple-crane lift – center of gravity location.

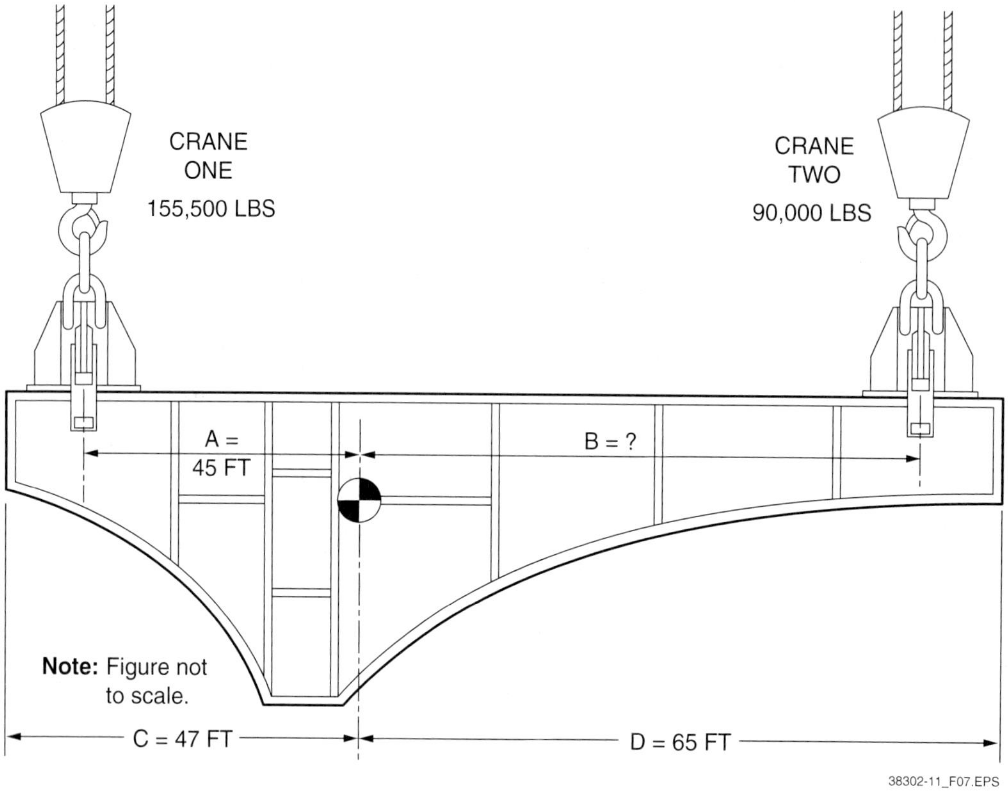

Figure 7 Multiple-crane lift – distance B, Trial One.

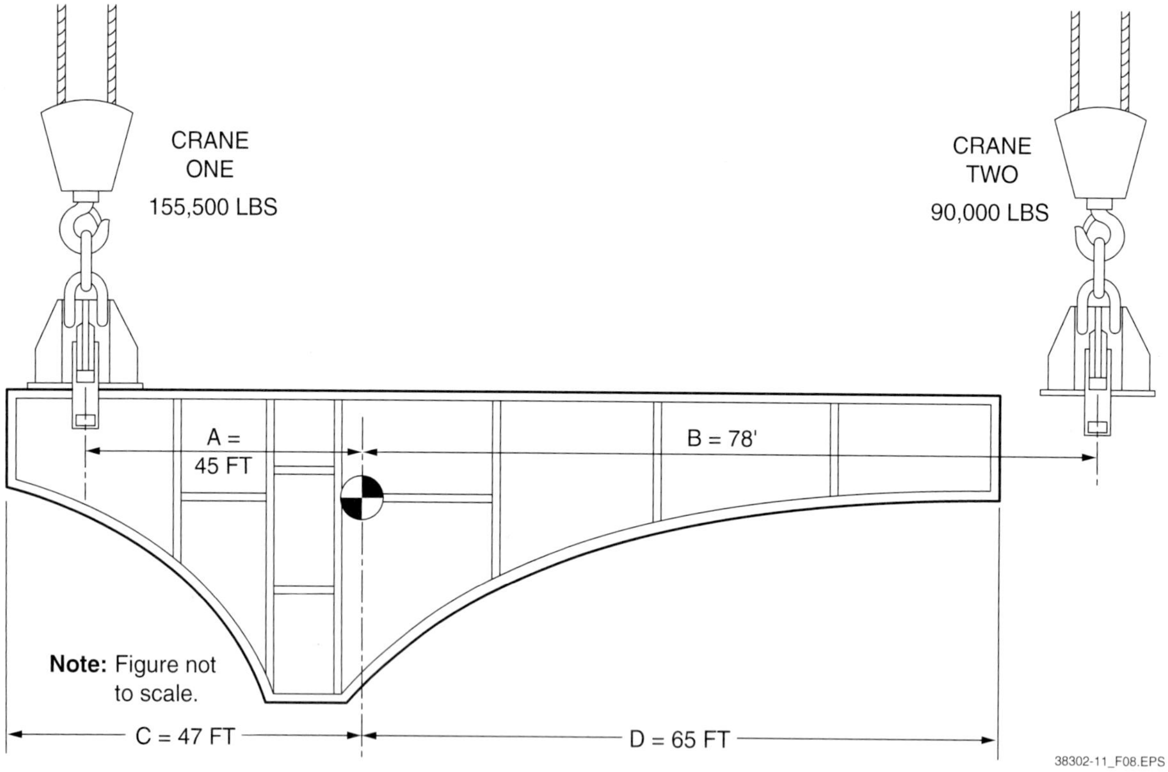

Figure 8 Multiple-crane lift – distance B, Trial One results.

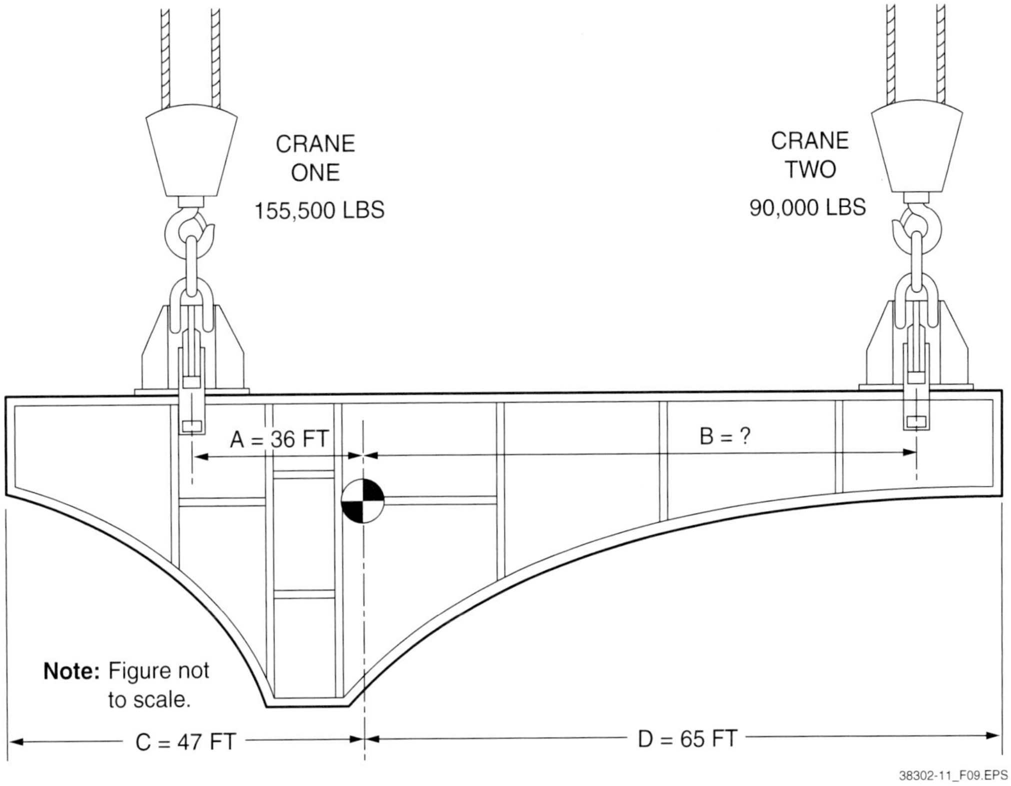

Figure 9 Multiple-crane lift – distance B, Trial Two.

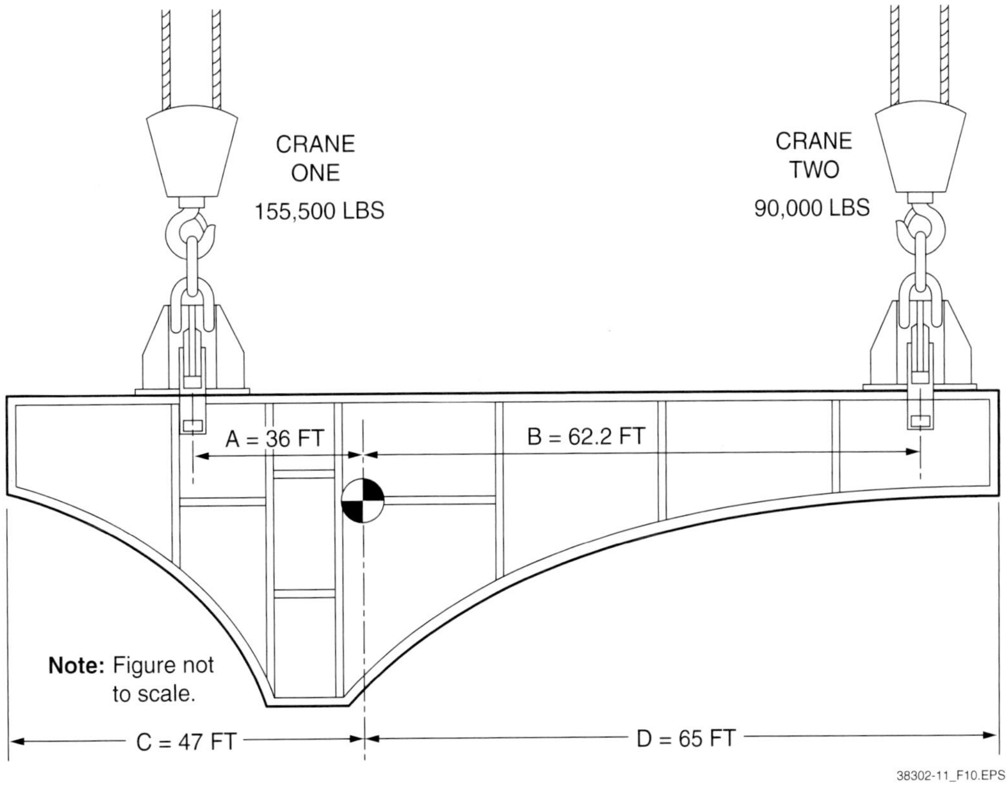

Figure 10 Multiple-crane lift – distance B, Trial Two results.

Step 7 Determine the actual load that each crane will be carrying:

$$\text{Load on Crane One} = \frac{B}{A + B} \times \text{load weight}$$

$$= \frac{62.2}{36 + 62.2} \times 180{,}000 \text{ lbs}$$

$$= 114{,}012.2 \text{ lbs}$$

This load is less than Crane One's rated net capacity of 155,500 lbs, and is under 75 percent (73.3 percent) of the capacity.

$$\text{Load on Crane Two} = \frac{A}{A + B} \times \text{load weight}$$

$$= \frac{36}{36 + 62.2} \times 180{,}000 \text{ lbs}$$

$$= 65{,}988 \text{ lbss}$$

This load is less than Crane Two's rated net capacity of 90,000 lbs and is under 75 percent (73.3 percent) of the capacity.

Step 8 Since neither crane is overloaded nor loaded beyond 75 percent of its capacity, the lift can proceed with A = 36 ft and B = 62.2 ft. Crane One will be 36 ft from the load's center of gravity, and Crane Two will be 62.2 ft from it.

NOTE

If the load radius of either crane is changed, repeat all of the steps using the new net capacity of the crane(s).

Multiple crane lifts can also use **equalizer** beams (*Figure 11*). Equalizer beams distribute the weight of a load between multiple cranes. By adjusting the crane and/or load attachment points on the beam, it is possible to share the load weight between the cranes in any proportion desired.

The operator can determine the location on the equalizer beam where the load is to be picked up, allowing each crane to share the load proportionately to its capacity, by following this method:

Step 1 Determine the lowest value of rated net capacity that Crane One will have when lifting from the main boom or an extension during the whole operation. Consider all factors, including configuration, operational quadrant, and load radius. Remember the load radius of each crane is measured to its pick-up point on the equalizer, not to the center of gravity of the load.

Step 2 Repeat Step 1 for Crane Two.

Step 3 Check to ensure that the sum of the lowest rated net capacities of the cranes exceeds the load weight plus the equalizer and sling weight. Ideally the combined weight should not exceed 75 percent of the sum of the net capacities.

Step 4 Measure the distance between the crane pick-up points on the equalizer beam (A).

Step 5 Establish the attachment point of the load on the equalizer beam according to the following formulas:

$$B = \frac{\text{net capacity of Crane Two}}{\text{total net capacity of Cranes One and Two}} \times A$$

$$C = \frac{\text{net capacity of Crane One}}{\text{total net capacity of Cranes One and Two}} \times A$$

Step 6 When the equalizer pick-up points have been established, determine the share of the load weight, including the weight of the equalizer carried by each crane:

$$\text{Load on Crane One} = \frac{C}{A} \times (\text{load weight})$$

$$\text{Load on Crane Two} = \frac{B}{A} \times (\text{load weight})$$

Check to ensure that neither crane is loaded beyond its lowest rated net capacity. Remember, the best situation is where neither crane is loaded in excess of 75 percent of its lowest rated net capacity. An example of this procedure follows (see *Figure 12*).

Step 1 Check the total capacity of the cranes:

Crane One = 120,000 lbs

Crane Two = 80,000 lbs

Crane One + Crane Two = total capacity

120,000 lbs + 80,000 lbs
= 200,000 lbs total capacity

This is greater than the total load weight of 143,000 lbs (load + equalizer). Also, 75 percent of the capacities' sum exceeds the load weight (75 percent of 200,000 = 150,000 lbs).

Step 2 Establish the load's pick-up point.

$$B = \frac{\text{net capacity of Crane Two}}{\text{total net capacity of Cranes One and Two}} \times A$$

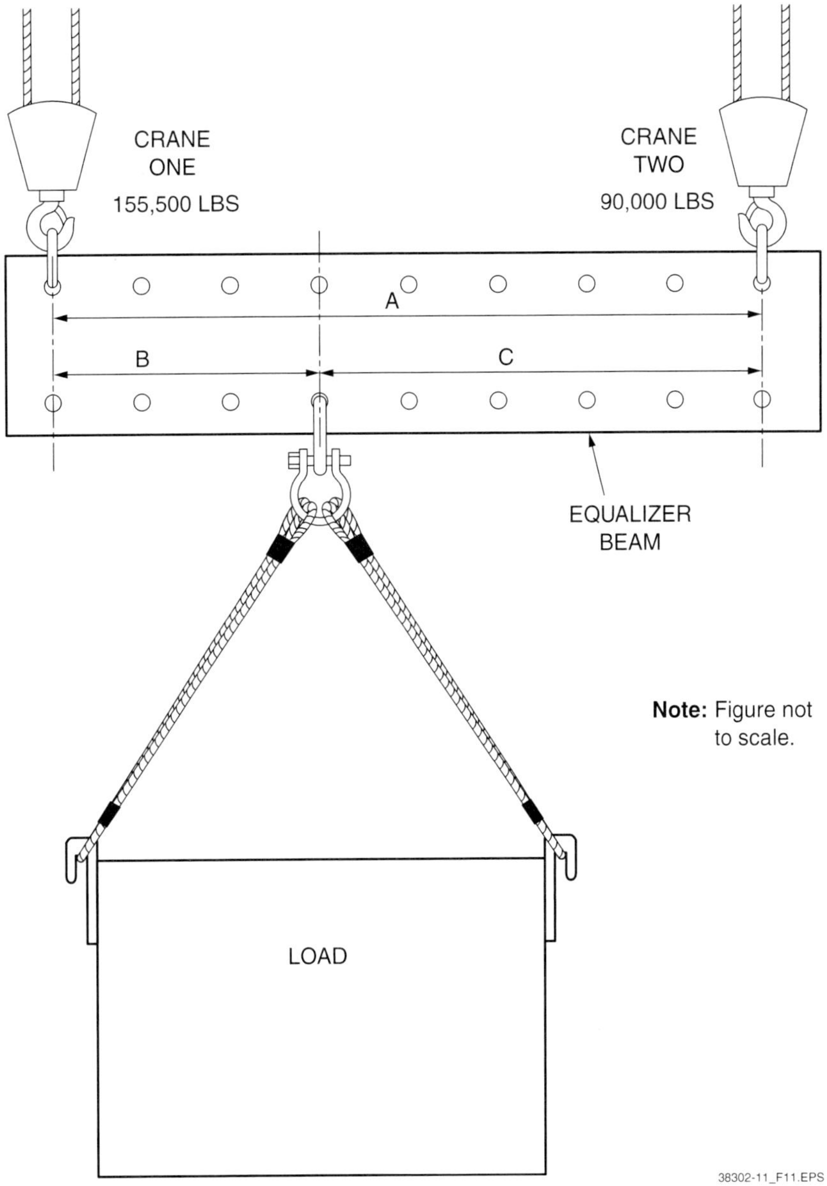

Figure 11 Multiple-crane lift with equalizer beam.

$$B = \frac{80{,}000 \text{ lbs}}{80{,}000 \text{ lbs} + 120{,}000 \text{ lbs}} \times A$$

$$B = \frac{80{,}000 \text{ lbs}}{200{,}000 \text{ lbs}} \times 25 \text{ ft}$$

B = 10 ft; therefore, C = 15 ft, as shown in *Figure 13*

Step 3 Determine the load that each crane will actually be carrying.

$$\text{Load on Crane One} = \frac{C}{A} \times (\text{total load weight})$$

$$= \frac{15}{25} \times 143{,}000 \text{ lbs}$$

$$= 85{,}800 \text{ lbs}$$

This number is less than the lowest rated net capacity and also meets the 75-percent capacity guideline (75 percent of 120,000 = 90,000 lbs).

$$\text{Load on Crane Two} = \frac{B}{A} \times (\text{load weight})$$

$$= \frac{10}{25} \times 143{,}000 \text{ lbs}$$

$$= 57{,}200 \text{ lbs}$$

This number is less than the lowest rated net capacity and also meets the 75 percent capacity guideline (75 percent of 80,000 = 60,000 lbs).

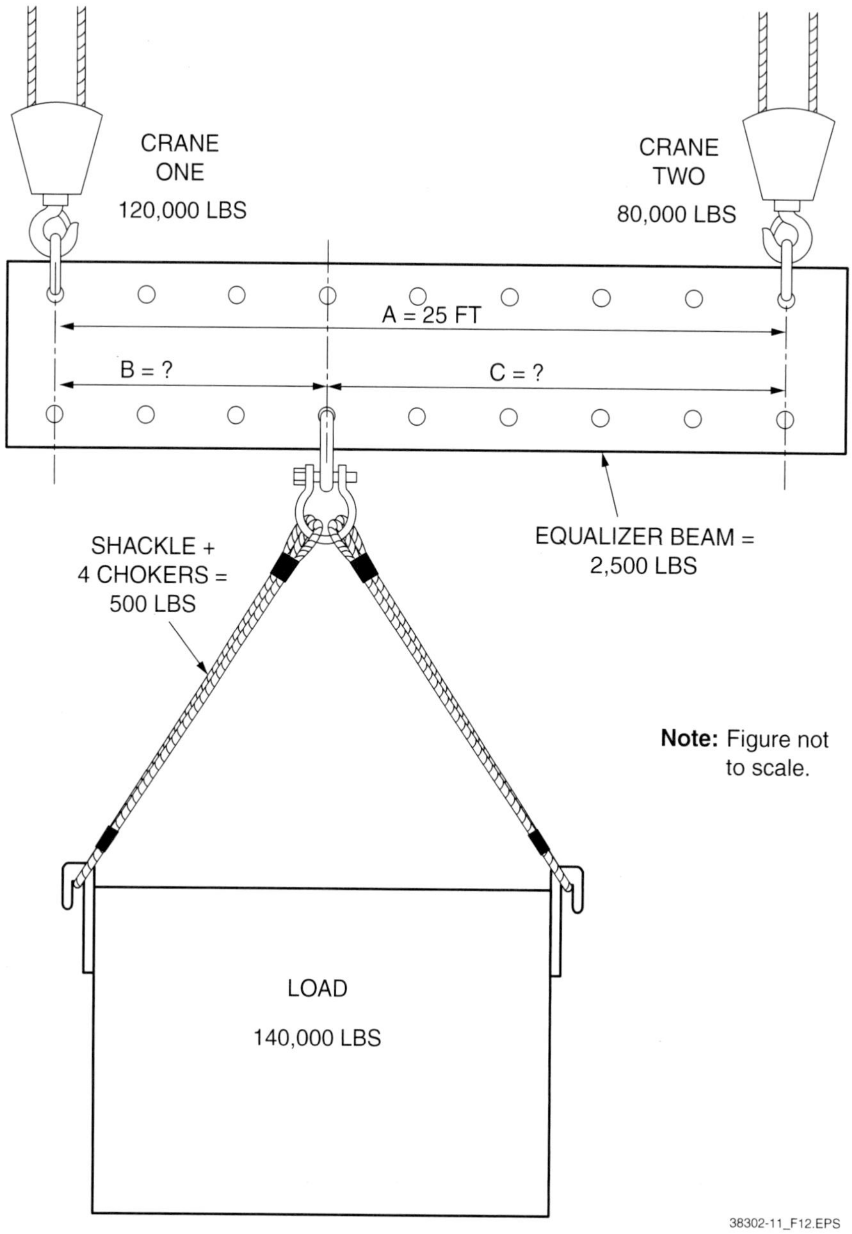

Figure 12 Multiple-crane lift with equalizer beam, determining distance B and C.

4.0.0 Crane Configuration

The load/capacity charts of mobile cranes are based on the configuration of the crane at the time of the lift. These configurations start with one of three basic situations:

- Jibs and boom extensions not installed
- Jibs and/or boom extensions installed but load lifted from main boom
- Jibs and/or boom extensions installed and load lifted from jib or boom extension

All other aspects of the crane's configuration are considered in relation to these three basic cases. The principal factors influencing the machine's capacity and how the chart is read include these considerations:

- Boom length, boom angle, and load radius
- Type of boom
  - Hydraulic
  - Lattice
- Operational quadrant in which the superstructure is working
- Configuration of crane base
  - On rubber
  - On outriggers (stabilizers)
  - On crawlers
- Hydraulic boom modes (A or B)
- Effective/stowed weight of jib or boom extension (or other attachments)

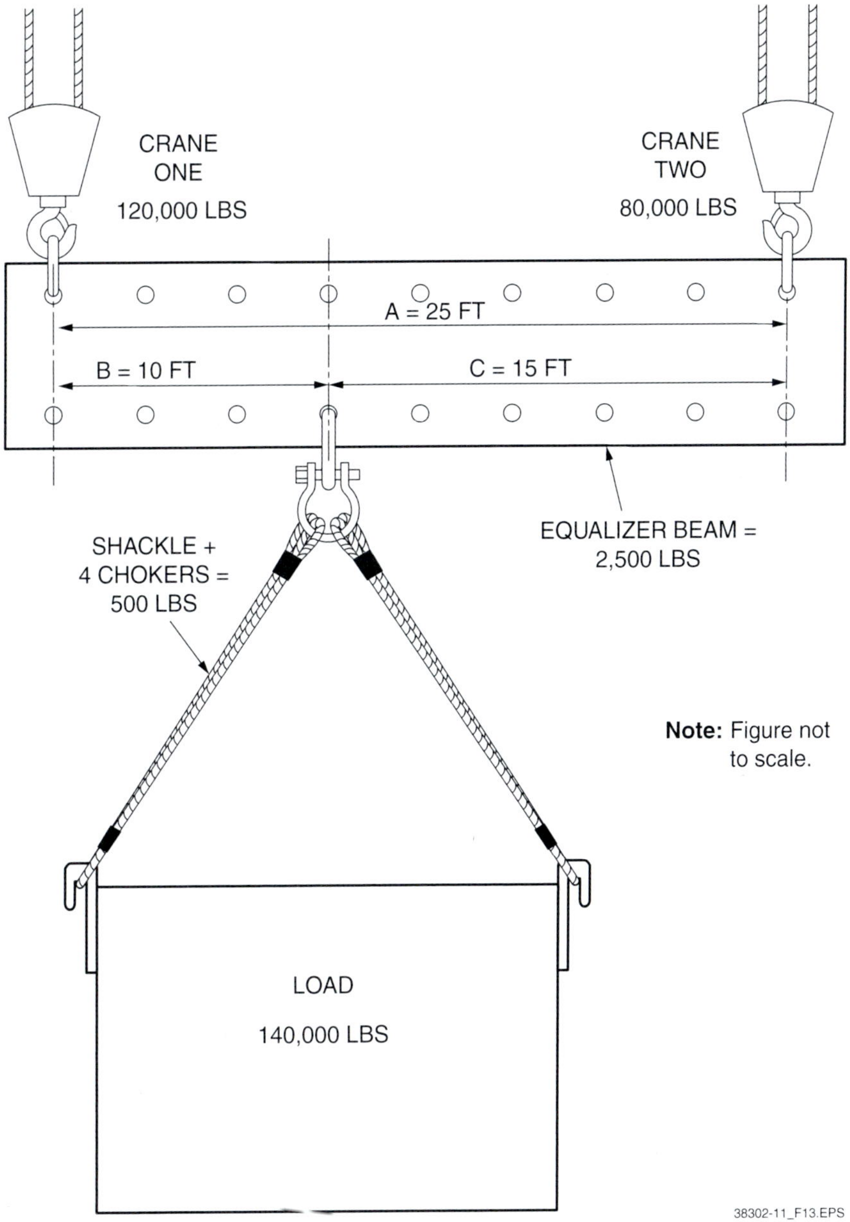

Figure 13 Multiple-crane lift with equalizer beam, distance B and C results.

- Ring or tower attachment
- Type, size, and weight of load block and headache ball
- Weight of all required rigging
- Weight of rope per manufacturer's requirement
- Counterweight configuration

4.1.0 Boom Length, Boom Angle, and Load Radius

For hydraulic (telescopic) cranes, the load radius is located on the left-hand side, and sometimes on the far right-hand side, of load/capacity charts. For lattice boom cranes, the load radius is usually located only on the left-hand side of the chart. The location of the boom angle information can vary. It may be in a column to the right of the load ra dius, or it may be set off in parentheses under the gross capacities.

Understanding where these terms are on a load/capacity chart is key to reading and interpreting a load/capacity chart properly. Operators must demonstrate their ability to read, comprehend, and write in the language of the crane manufacturer's operation and maintenance instruction materials. They must also exhibit arithmetic skills and the ability to use the manufacturer's load/capacity charts.

Most hydraulic load/capacity charts (*Figure 14*) start with the shortest boom length at the upper left side and read to the right and down for progressively longer booms. First locate the number

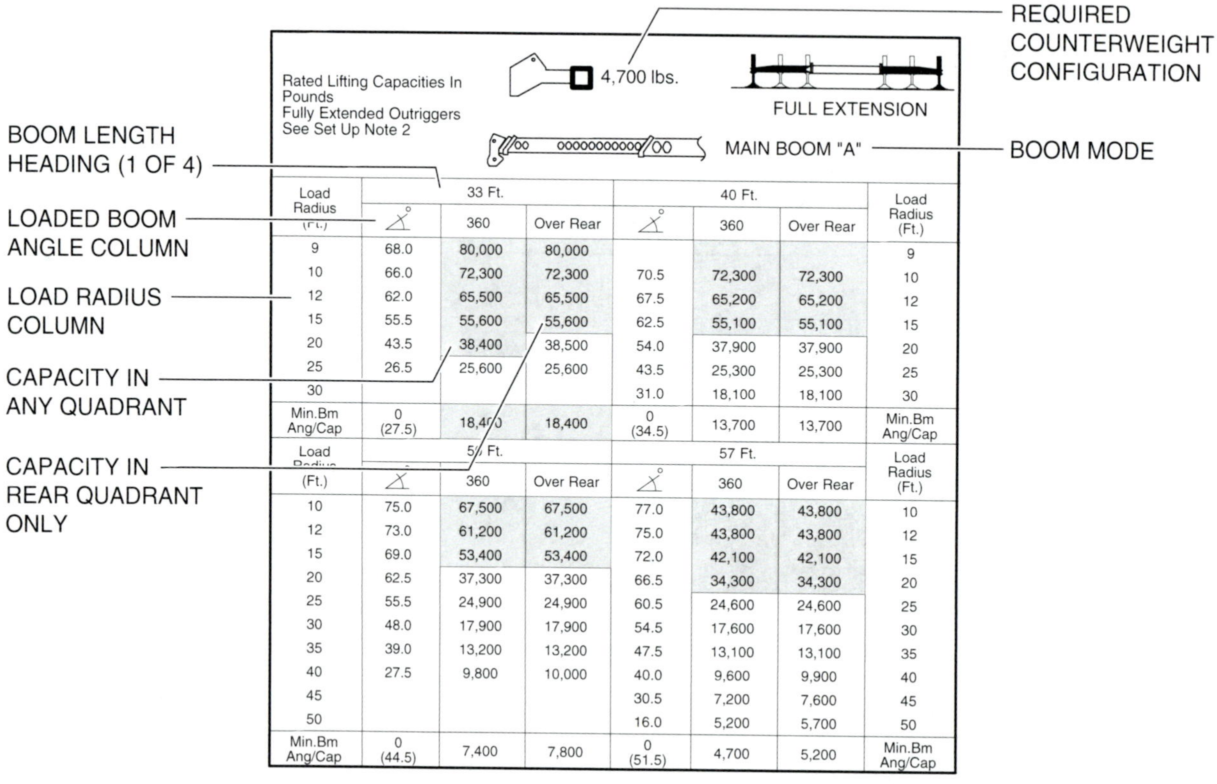

Load Radius (Ft.)	33 Ft.			40 Ft.			Load Radius (Ft.)
	∠°	360	Over Rear	∠°	360	Over Rear	
9	68.0	80,000	80,000				9
10	66.0	72,300	72,300	70.5	72,300	72,300	10
12	62.0	65,500	65,500	67.5	65,200	65,200	12
15	55.5	55,600	55,600	62.5	55,100	55,100	15
20	43.5	38,400	38,500	54.0	37,900	37,900	20
25	26.5	25,600	25,600	43.5	25,300	25,300	25
30				31.0	18,100	18,100	30
Min.Bm Ang/Cap	0 (27.5)	18,400	18,400	0 (34.5)	13,700	13,700	Min.Bm Ang/Cap
Load Radius (Ft.)	51 Ft.			57 Ft.			Load Radius (Ft.)
	∠°	360	Over Rear	∠°	360	Over Rear	
10	75.0	67,500	67,500	77.0	43,800	43,800	10
12	73.0	61,200	61,200	75.0	43,800	43,800	12
15	69.0	53,400	53,400	72.0	42,100	42,100	15
20	62.5	37,300	37,300	66.5	34,300	34,300	20
25	55.5	24,900	24,900	60.5	24,600	24,600	25
30	48.0	17,900	17,900	54.5	17,600	17,600	30
35	39.0	13,200	13,200	47.5	13,100	13,100	35
40	27.5	9,800	10,000	40.0	9,600	9,900	40
45				30.5	7,200	7,600	45
50				16.0	5,200	5,700	50
Min.Bm Ang/Cap	0 (44.5)	7,400	7,800	0 (51.5)	4,700	5,200	Min.Bm Ang/Cap

Note: Refer to page 5 for (Capacity Deductions for Auxiliary Load Handling Equipment).

∠° Loaded Boom Angle in Degrees

( ) Reference radius for minimum boom angle capacities (shown in parenthesis) are in feet.

38302-11_F14.EPS

Figure 14 Hydraulic boom load/capacity chart.

that corresponds to the length in feet of the boom under the boom length heading. Then, locate the operating radius of the crane in the left- or right-hand column. The capacity where the operating radius line intersects with a particular operating quadrant under the boom length column is the gross capacity that can be lifted safely with a defined counterweight configuration.

Lattice boom crawler crane load/capacity charts (*Figure 15*) can be presented differently, depending on the manufacturer. In this example, the boom length is located in the chart heading, and the load radius is in the left-hand column.

The operator must first know what boom length the crane is operating at, determine t he radius, and then read across to determine the gross capacity, depending on the counterweight configuration. Some manufacturers may also include the **boom point elevation** in their charts for reference.

Remember that the load/capacity charts shown in this module are examples from certain manufacturers. The standards do not require or specify that this information be presented in a particular format. *ASME B30.5* simply requires that load charts contain certain specific information.

Load/capacity chart layouts are as diverse as the number of crane manufacturers. They can even vary greatly between different crane models from the same manufacturer. Some have illustrations and counterweight requirements at the top; some express their load radius or boom lengths in feet and/or meters (*Figures 16, 17,* and *18*), and some list their capacities in kilograms and pounds (*Figures 16* and *18*). This variety presents a challenge to operators; they must pay close attention to load/capacity charts and their introductory information when switching between cranes to avoid making mistakes on load chart calculations and interpretations.

It is imperative for operators to remember that if a load radius, boom length, or boom angle falls between values listed on the load/capacity chart, there is no **interpolation** of values allowed. When either boom length or radius, or both, are between values listed, the smallest load shown at either the next larger radius or next longer boom length shall be used.

CAUTION

Interpolation (mathematical approximation) is not permitted in determining load chart capacities.

BOOM LENGTH (CHART HEADING)

LOAD RADIUS COLUMN

CAPACITY IN ANY QUADRANT

LOADED BOOM ANGLE COLUMN

CAPACITY IN EITHER END QUADRANT ONLY

COUNTERWEIGHT CONFIGURATIONS (LWR = BOTH LOWER CRAWLER FRAME COUNTERWEIGHTS)

Main Boom Capacities* – 110 Ft. — Open Throat Boom

Load Radius (ft)	Boom Angle (deg)	360° Rotation				Over End Blocked	
		135K + LWR (lb)	135K + 0 (lb)	83K + 0 (lb)	31K + 0 (lb)	135K + LWR(lb)	135K + 0 (lb)
18.5	82.0	346,500	346,500	329,000	283,900	346,500	346,500
19	81.7	344,400	344,400	320,000	276,000	344,400	344,400
20	81.2	341,100	341,100	303,900	246,900	341,100	341,100
25	78.6	276,100	276,100	233,200	158,900	276,100	276,100
30	75.9	229,100	226,500	171,200	116,000	229,100	229,100
35	73.2	192,600	178,500	134,500	90,500	195,200	195,200
40	70.4	158,500	146,800	110,300	73,700	169,700	169,700
50	64.8	116,200	107,400	80,100	52,900	133,700	133,000
60	58.9	90,900	83,900	62,200	40,400	109,700	102,800
70	52.6	74,100	68,300	50,200	32,100	90,700	83,100
80	45.7	62,100	57,200	41,600	26,100	75,800	69,300
90	37.9	53,100	48,800	35,200	21,700	64,700	59,100
100	28.3	46,100	42,200	30,200	18,100	56,100	51,100
110	13.8	40,400	36,900	26,100	15,200	49,200	44,700

*Ratings are not more than 75% of tipping loads

38302-11_F15.EPS

Figure 15 Lattice boom crawler crane load/capacity chart.

Outriggers Fully Extended

25' - 43' *(7.62m - 13.11m)* Offset Telescoping Fly + 70.25' *(21.41m)* Main Boom with Manual Retracted

LOAD RADIUS (ft/m)	25' (7.62m) Fly: 2° OFFSET LOADED BOOM ANGLE (deg.)	2° OFFSET LOAD (lbs/kgs)	15° OFFSET LOADED BOOM ANGLE (deg.)	15° OFFSET LOAD (lbs/kgs)	30° OFFSET LOADED BOOM ANGLE (deg.)	30° OFFSET LOAD (lbs/kgs)	43' (13.11m) Fly: 2° OFFSET LOADED BOOM ANGLE (deg.)	2° OFFSET LOAD (lbs/kgs)	15° OFFSET LOADED BOOM ANGLE (deg.)	15° OFFSET LOAD (lbs/kgs)	30° OFFSET LOADED BOOM ANGLE (deg.)	30° OFFSET LOAD (lbs/kgs)	LOAD RADIUS (ft/m)
20 *6.10*	78.5	12,700 *5,761*											20 *6.10*
25 *7.62*	75.5	11,500 *5,216*	79.0	8,800 *3,992*			77.5	4,900 *2,223*					25 *7.62*
30 *9.14*	72.5	10,600 *4,808*	76.0	8,200 *3,720*	79.0	6,100 *2,767*	75.5	4,700 *2,132*					30 *9.14*
35 *10.67*	69.5	9,600 *4,355*	73.0	7,500 *3,402*	76.0	5,700 *2,586*	73.0	4,500 *2,041*	78.0	4,000 *1,614*			35 *10.67*
40 *12.19*	66.0	8,700 *3,946*	69.5	6,900 *3,130*	72.5	5,400 *2,449*	70.0	4,300 *1,950*	75.5	3,900 *1,769*			40 *12.19*
45 *13.72*	63.0	8,000 *3,629*	66.0	6,400 *2,903*	69.0	5,100 *2,313*	67.5	4,100 *1,860*	72.5	3,700 *1,678*	77.5	3,000 *1,361*	45 *13.72*
50 *15.24*	59.5	7,400 *3,357*	62.5	5,900 *2,676*	65.5	4,800 *2,177*	65.0	3,900 *1,769*	70.0	3,500 *1,588*	74.5	2,900 *1,315*	50 *15.24*
55 *16.76*	56.0	6,500 *2,948*	59.0	5,300 *2,404*	62.0	4,500 *2,041*	62.0	3,800 *1,724*	67.0	3,300 *1,497*	71.5	2,700 *1,225*	55 *16.76*
60 *18.29*	52.0	5,400 *2,449*	55.0	4,900 *2,223*	58.0	4,200 *1,905*	59.5	3,700 *1,678*	64.0	3,100 *1,406*	68.5	2,600 *1,179*	60 *18.29*
65 *19.81*	48.0	4,500 *2,041*	51.0	4,500 *2,041*	53.5	3,900 *1,769*	56.5	3,500 *1,588*	61.0	2,900 *1,315*	65.5	2,500 *1,134*	65 *19.81*
70 *21.34*	43.5	3,700 *1,678*	46.5	3,900 *1,769*	49.0	3,600 *1,633*	53.0	3,200 *1,452*	58.0	2,800 *1,270*	62.5	2,400 *1,089*	70 *21.34*
75 *22.86*	38.5	3,000 *1,361*	41.5	3,200 *1,452*	44.0	3,400 *1,542*	50.5	3,000 *1,361*	54.5	2,600 *1,179*	59.0	2,300 *1,043*	75 *22.86*
80 *24.38*	33.0	2,500 *1,134*	36.0	2,600 *1,179*	37.5	2,700 *1,225*	46.5	2,800 *1,270*	51.0	2,500 *1,134*	55.0	2,200 *998*	80 *24.38*
85 *25.91*	26.5	2,000 *907*	29.0	2,100 *953*	30.0	2,100 *953*	43.0	2,700 *1,225*	47.5	2,400 *1,089*	51.0	2,100 *953*	85 *25.91*
90 *27.43*	17.5	1,600 *726*	19.5	1,600 *726*	17.0	1,600 *726*	38.5	2,300 *1,043*	43.0	2,300 *1,043*	46.5	2,000 *907*	90 *27.43*
95 *28.96*							34.0	2,000 *907*	38.5	2,200 *998*	41.5	2,000 *907*	95 *28.96*
100 *30.48*							28.5	1,600 *726*	33.0	1,800 *816*	35.5	1,900 *862*	100 *30.48*
105 *32.00*									25.5	1,400 *635*	26.0	1,400 *635*	105 *32.00*

Warning: Do not lower 25' offset telescoping fly in working position below 12∞ unless main boom is 68' or less, since loss of stability will occur causing a tipping condition.

Warning: Do not lower 43' offset telescoping fly in working position below 20∞ unless main boom is 65' or less, since loss of stability will occur causing a tipping condition.

38302-11_F16.EPS

Figure 16 Load/capacity chart in English and metric units with left- and right-hand load radius columns.

Lifting capacities main boom
Tragfähigkeiten Hauptausleger
Capacités de levage flèche principale

140 t + 40 t ZB — 7,25 m — 360° — **DIN/ISO**

SH

Radius / Ausladung / Porteé (m)	18,0 m	24,0 m	30,0 m	36,0 m	42,0 m	48,0 m	54,0 m	60,0 m	66,0 m	72,0 m	78,0 m	84,0 m
m	t	t	t	t	t	t	t	t	t	t	t	t
4,5	300,0*	295,0*	-	-	-	-	-	-	-	-	-	-
5	300,0*	295,0*	289,0*	-	-	-	-	-	-	-	-	-
5,5	300,0*	295,0*	289,0*	274,0*	245,0	-	-	-	-	-	-	-
6	300,0*	295,0*	289,0*	274,0*	245,0	215,0	-	-	-	-	-	-
7	288,0*	295,0*	280,0*	274,0*	245,0	215,0	171,0	143,0	116,0	97,5	-	-
8	271,0*	276,0*	271,0*	255,0*	245,0	215,0	171,0	143,0	116,0	97,5	82,5	69,5
9	251,0*	247,0	249,0	235,0	225,0	202,0	171,0	143,0	116,0	97,5	82,5	69,5
10	209,0	209,0	209,0	208,0	205,0	189,0	165,0	139,0	116,0	97,5	82,5	69,5
12	157,0	156,0	156,0	155,0	155,0	154,0	151,0	133,0	115,0	97,0	82,0	69,5
14	125,0	124,0	124,0	123,0	122,0	122,0	122,0	121,0	113,0	96,5	81,5	69,0
16	103,0	103,0	102,0	101,0	101,0	100,0	100,0	100,0	100,0	96,0	81,5	68,5
18	88,5	87,5	87,0	86,0	85,5	85,0	84,5	84,5	84,0	83,5	81,0	68,5
20	-	76,0	75,0	74,5	73,5	73,0	73,0	72,5	72,5	72,0	71,0	66,5
22	-	67,0	66,0	65,5	64,5	64,0	63,5	63,5	63,0	62,5	62,0	61,5
24	-	-	59,0	58,0	57,5	56,5	56,5	56,0	56,0	55,0	54,5	54,0
26	-	-	53,0	52,0	51,5	50,5	50,5	50,0	49,9	49,4	48,7	48,0
28	-	-	48,3	47,4	46,5	45,9	45,5	45,1	44,9	44,3	43,6	42,9
30	-	-	-	43,2	42,3	41,7	41,2	40,8	40,6	40,0	39,4	38,6
34	-	-	-	-	35,7	35,0	34,4	34,0	33,8	33,2	32,5	31,8
38	-	-	-	-	30,7	29,9	29,3	28,8	28,5	27,9	27,2	26,5
42	-	-	-	-	-	25,9	25,3	24,7	24,4	23,8	23,1	22,4
46	-	-	-	-	-	-	22,1	21,5	21,1	20,5	19,7	18,8
50	-	-	-	-	-	-	-	18,8	18,4	17,6	16,8	15,9
54	-	-	-	-	-	-	-	16,5	16,0	15,2	14,4	13,5
58	-	-	-	-	-	-	-	-	14,0	13,2	12,3	11,4
62	-	-	-	-	-	-	-	-	-	11,5	10,6	9,6
66	-	-	-	-	-	-	-	-	-	-	9,1	8,1
70	-	-	-	-	-	-	-	-	-	-	7,8	6,8
74	-	-	-	-	-	-	-	-	-	-	-	5,7

Main boom · Hauptausleger · Flèche principale

38302-11_F17.EPS

Figure 17 Load/capacity chart in metric units only with left-hand radius column.

Some hydraulic telescoping cranes are equipped with dual boom mode capabilities, known as A and B (*Figures 19* and *20*). They are engaged or disengaged with a switch. Boom mode A could extend all sections of the boom equally, whereas boom mode B could just extend the first, second, or third section. The different configurations will result in different lifting capacities, which, in turn, necessitate different load/capacity charts for each mode.

4.2.0 Quadrants of Operation

As a crane swings to various quadrants, the distance between the load's position and the tipping axis changes dramatically. This is because cranes are typically built to be longer than they are wide to allow them to fit on a roadway or trailer for transportation from job to job. This rectangular shape tends to result in more stability of the crane from front to rear than from side to side.

The operational quadrants are termed *over-the-front*, *over-the-rear*, *over-the-side*, and *360 degrees* (*Figure 21*).

Crane manufacturers produce separate load/capacity charts for certain quadrants because capacities will vary. For example, notice the capacity differences in the load/capacity charts in *Figure 22* for a hydraulic boom crane.

With a boom length of 36 feet, the gross capacity remains the same in both configurations at an operating radius of 10 feet. However, at all other boom lengths and operating radii, the capacity for 360 degrees is less than for over-the-front operation. The differences in capacities can vary significantly at different radii as the boom length increases.

4.3.0 Configuration of the Crane Base

The crane can be set up on rubber, outriggers, with crawlers extended or retracted, or on stabilizers. Each configuration affects capacity, which translates into a different load/capacity chart.

TEREX LIFTING

AMERICAN HC 60

Hydraulic Crawler Crane 46 HI Boom

With 46HI Angle Boom, 4 Sheave Tip, 39,000 Pound Counterweight

Boom Length	Radius (Feet)	Boom Angle (Degrees)	Side Frames Extended (Pounds)	From Boom Pt. to Ground (Feet)
40' (12.2M)	10	80.3	120,000*	45
	12	77.4	120,000*	44
	15	73.0	93,390	44
	20	65.3	59,020	42
	25	57.1	42,780	39
	30	48.1	33,370	35
	35	37.5	27,190	30
	40	23.4	22,840	21
50' (15.2M)	12	80.0	120,000*	55
	15	76.4	93,310	54
	20	70.5	58,910	53
	25	64.2	42,640	50
	30	57.7	33,230	48
	35	50.6	27,040	44
	40	42.7	22,670	39
	45	33.5	19,460	33
	50	20.9	16,960	23
60' (18.3M)	13	80.7	117,460*	65
	15	78.7	93,230	64
	20	73.8	58,800	63
	25	68.8	42,510	61
	30	63.6	33,100	59
	35	58.1	26,890	56
	40	52.3	22,520	53
	45	46.0	19,330	49
	50	38.9	16,820	43
	55	30.5	14,820	36
	60	19.0	13,190	[illegible]

Boom Length	Radius (Feet)	Boom Angle (Degrees)	Side Frames Extended (Pounds)	From Boom Pt. to Ground (Feet)
90' (con't)	45	62.4	18,820	85
	50	58.7	16,290	82
	55	54.9	14,290	79
	60	50.9	12,650	75
	65	46.7	11,300	71
	70	42.2	10,160	66
	75	37.2	9,180	60
	80	31.5	8,330	52
	85	24.8	7,610	43
	90	15.5	6,970	29
100' (30.5M)	19	81.0	62,850	104
	20	80.4	58,180	104
	25	77.5	41,800	103
	30	74.5	32,400	102
	35	71.5	26,170	100
	40	68.5	21,770	98
	45	65.3	18,630	96
	50	62.2	16,110	94
	55	58.9	14,100	91
	60	55.4	12,460	88
	65	51.9	11,110	84
	70	48.2	9,970	80
	75	44.2	8,980	75
	80	39.9	8,140	[illegible]
	85	35.2	[illegible]	[illegible]
	90	[illegible]	[illegible]	[illegible]

Boom Length	Radius (Feet)	Boom Angle (Degrees)	Side Frames Extended (Pounds)	From Boom Pt. to Ground (Feet)
120' (con't)	90	43.7	6,350	88
	95	40.2	5,770	83
	100	36.3	5,260	76
	105	32.0	4,800	69
	110	27.2	4,380	60
	115	21.4	4,020	49
	120	13.4	3,680	33
130' (39.6M)	24	80.8	43,780	134
	25	80.4	41,220	134
	30	78.1	31,830	133
	35	75.9	25,570	131
	40	73.6	21,160	130
	45	71.3	18,050	129
	50	68.9	15,520	127
	55	66.6	13,500	125
	60	64.1	11,870	122
	65	61.7	10,500	120
	70	59.1	9,350	117
	75	56.5	8,370	114
	80	53.8	7,520	110
	85	51.0	6,780	107
	90	[illegible]	[illegible]	102

38302-11_F18.EPS

Figure 18 Load/capacity chart in English and metric units with left-hand boom length column.

4.3.1 On Rubber

On-rubber load/capacity charts are shown in *Figure 23*. Depending on the manufacturer, the required tire pressures may be defined in notes for the charts or in a separate table.

There are several requirements for on-rubber lifting. Capacities for on-rubber lifting do not exceed 75 percent of tipping loads as determined by test in accordance with *ASME/SAE J765*. This percentage is strictly an **engineering** stipulation used by the manufacturer in creating load/capacity chart capacities. The operator does not have a 25-percent cushion to work with, because the load chart capacities must not be exceeded. Note that the crane capacity must be reduced if the tires are under-inflated (Note 9 on *Figure 23*).

4.3.2 On Outriggers

Capacities for on-outrigger lifting do not exceed 85 percent of tipping loads. As mentioned with on-rubber lifting, this is an engineering stipulation. It does not mean the operator has a 15-percent cushion. Depending on the design of the crane, there may be load/capacity charts for outriggers at no extension, intermediate extension, and full extension (*Figures 24, 25,* and *26*). At no extension, capacities do not exceed 75 percent of tipping. In this situation, the crane may have less capacity but more stability than when on rubber.

35 - 138 ft. (10.8 - 42.0 m) | 0 lbs. (0 kg) | 100% | 360°

85% Domestic (Pounds)

(Feet)	35	61	74	87	99	112	138
10	139,500 (65.5)	42,900 (77.5)	*32,100 (80)				
12	110,000 (62)	42,000 (75.5)	32,100 (78.5)	*31,850 (80)			
15	92,450 (56)	36,550 (72.5)	32,100 (76)	31,850 (78.5)	*21,350 (80)		
20	64,600 (44.5)	29,400 (67.5)	31,350 (71.5)	28,850 (75)	21,350 (77.5)	19,000 (79.5)	
25	48,350 (29.5)	24,350 (62)	26,450 (67.5)	25,050 (71.5)	20,850 (74.5)	18,150 (77)	*19,000 (80)
30		20,500 (56.5)	22,300 (63)	21,550 (68)	18,650 (71.5)	17,300 (74.5)	18,300 (78.5)
35		17,450 (50)	19,100 (58.5)	18,500 (64)	16,900 (68.5)	16,450 (71.5)	17,650 (76.5)
40		15,050 (43)	16,500 (53.5)	16,000 (60)	15,300 (65)	15,650 (69)	17,000 (74)
45		13,100 (35)	14,450 (48.5)	14,000 (56)	13,650 (61.5)	14,150 (66)	16,350 (72)
50		11,450 (24.5)	12,750 (42.5)	12,350 (52)	12,100 (58.5)	12,700 (63)	15,450 (69.5)
60			9,160 (28)	9,710 (42.5)	9,580 (51)	10,150 (57)	10,400 (65)
70				5,990 (30)	7,430 (42.5)	8,220 (50)	7,140 (60)
80					4,720 (32)	5,790 (42.5)	4,820 (55)
90					2,550 (15.5)	3,700 (33.5)	3,100 (49.5)
100						1,990 (21)	1,760 (43)
Minimum boom angle (deg.) for indicated length (no load)						0	40

Maximum boom length (ft.) at 0 degree boom angle (no load) 112

NOTE: ( ) Boom angles are in degrees.

*This capacity is based on maximum boom angle.

Boom Angle	35	61	74	87	99	112
0	26,400 (28.2)	10,150 (53.8)	6,240 (66.6)	3,330 (79.4)	2,130 (92.2)	1,260 (105)

NOTE: ( ) Reference radii are in feet.

A6-829-014471

T1 T2 T3 T4 %

MODE A

38302-11_F19.EPS

Figure 19 Hydraulic crane load/capacity chart for mode A boom configuration.

35 - 138 ft.
(10.8 - 42.0 m)

0 lbs.
(0 kg)

100%

360°

85% Domestic (Pounds)

(Feet)	35	55	61	74	87	99	112	125	138
10	139,500 (65.5)	79,100 (76)	78,450 (77.5)	*57,050 (80)					
12	110,000 (62)	79,100 (73.5)	77,500 (75.5)	57,050 (78.5)	*43,300 (80)				
15	92,450 (56)	79,100 (70)	69,850 (72.5)	51,650 (76)	43,300 (78.5)	*32,100 (80)			
20	64,600 (44.5)	64,250 (64.5)	59,850 (67.5)	44,350 (71.5)	39,550 (75)	32,100 (77.5)	30,050 (79.5)	*20,150 (80)	
25	48,350 (29.5)	48,050 (58)	48,450 (62)	38,750 (67.5)	33,800 (71.5)	32,100 (74.5)	30,050 (77)	20,150 (79)	*19,000 (80)
30		34,400 (51)	34,050 (56.5)	34,050 (63)	29,200 (68)	30,200 (71.5)	27,350 (74.5)	19,100 (76.5)	18,300 (78.5)
35		25,150 (43.5)	24,500 (50)	25,200 (58.5)	25,250 (64)	26,600 (68.5)	24,300 (71.5)	18,100 (74)	17,650 (76.5)
40		18,650 (34.5)	18,400 (43)	19,300 (53.5)	18,750 (60)	20,050 (65)	21,350 (69)	17,250 (72)	17,000 (74)
45		14,150 (21.5)	14,150 (35)	14,400 (48.5)	14,100 (56)	15,350 (61.5)	16,550 (66)	16,450 (69)	16,350 (72)
50			11,050 (24.5)	10,750 (42.5)	10,700 (52)	11,850 (58.5)	13,050 (63)	14,250 (66.5)	15,450 (69.5)
60				5,810 (28)	6,040 (42.5)	7,110 (51)	8,210 (57)	9,310 (61.5)	10,400 (65)
70					2,970 (30)	3,990 (42.5)	5,030 (50)	6,080 (56)	7,140 (60)
80						1,780 (32)	2,780 (42.5)	3,800 (49.5)	4,820 (55)
90							1,120 (33.5)	2,100 (43)	3,100 (49.5)
100									1,760 (43)
Minimum boom angle (deg.) for indicated length (no load)					20	25	33	37	40
Maximum boom length (ft.) at 0 degree boom angle (no load)							74		

NOTE: ( ) Boom angles are in degrees.

*This capacity is based on maximum boom angle.

Boom Angle	35	55	61	74
0	26,400 (28.2)	12,500 (47.4)	9,190 (53.8)	3,540 (66.6)

NOTE: ( ) Reference radii are in feet.

A6-829-014539

T1 T2 T3 T4 %

MODE B

38302-11_F20.EPS

Figure 20 Hydraulic crane load/capacity chart for mode B boom configuration.

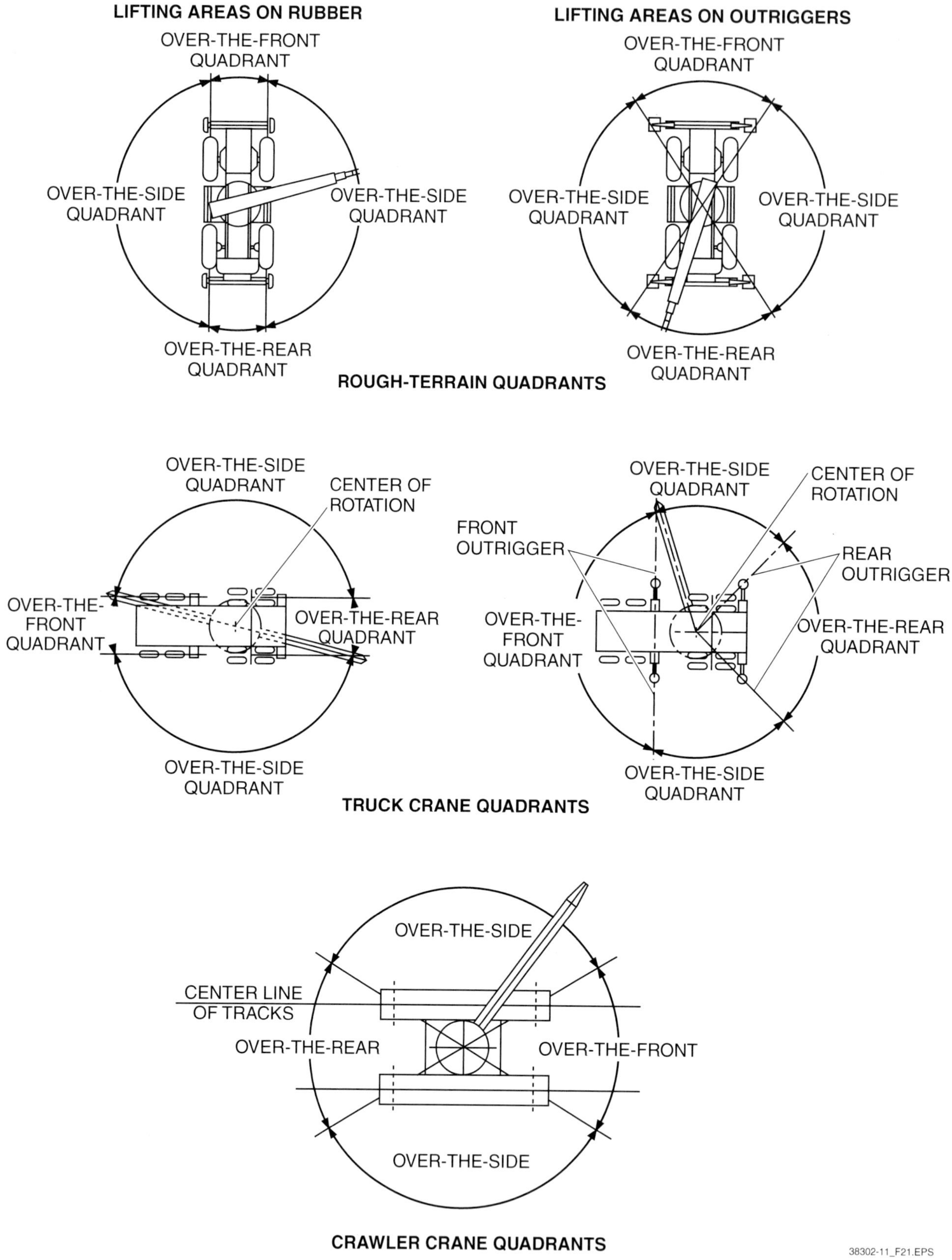

Figure 21 Quadrant definition drawings for rough-terrain, truck, and crawler cranes.

QUADRANT

36-70 ft. | 12,150 lbs | Stationary | 360°

Pounds

#9005

Feet	Main Boom Length in Feet 36	40	50	*60	70
10	45,300 (69)	39,700 (72)			
12	41,750 (65.5)	39,700 (68.5)	29,600 (73.5)		
15	29,350 (59.5)	26,450 (63.5)	26,450 (70)	20,900 (74)	
20	17,800 (49)	17,650 (55)	17,050 (63.5)	16,250 (69)	16,250 (72)
25	11,750 (36)	11,700 (45)	11,350 (56.5)	10,850 (63.5)	10,850 (68)
30		8,040 (31.5)	7,820 (48.5)	7,470 (57.5)	7,470 (63)
35			5,400 (40)	5,120 (51.5)	5,120 (58)
40			3,660 (28)	3,200 (45)	3,430 (53)
45				1,000 (37)	2,150 (47.5)
50					1,150 (41.0)

Lifting Capacities at Zero Degree Boom Angle On Rubber - Stationary 360

Boom Angle	Main Boom Length in Feet 36	40	50
0°	8,180 (29.7)	5,890 (34.2)	2,170 (44.2)

Note: ( ) Reference radii in feet. A6-829-101048A

#LMI operating code. Refer to LMI manual for instructions.
*60 ft. boom length is with inner-mid extended and outer-mid & fly retracted.

QUADRANT

36-70 ft. | 12,150 lbs | Stationary | Defined Arc Over Front

Pounds

#9005

Feet	Main Boom Length in Feet 36	40	50	*60	70
10	45,300 (69)	42,850 (72)	29,600 (76)		
12	43,650 (65.5)	41,350 (68.5)	29,600 (73.5)		
15	38,300 (59.5)	36,300 (63.5)	29,600 (70)	20,900 (74)	17,300 (76.5)
20	31,150 (49)	29,550 (55)	25,900 (63.5)	20,900 (69)	17,300 (72)
25	24,100 (36)	24,150 (45)	21,800 (56.5)	18,800 (63.5)	17,300 (68)
30		17,400 (31.5)	17,200 (48.5)	15,300 (57.5)	15,300 (63)
35			12,800 (40)	12,500 (51.5)	11,000 (58)
40			9,720 (28)	9,390 (45)	9,390 (53)
45				7,090 (37)	7,090 (47.5)
50				5,310 (26.5)	5,310 (41)
55					3,870 (33.5)
60					3,170 (24)

Lifting Capacities at Zero Degree Boom Angle On Rubber - Defined Arc Over Front

Boom Angle	Main Boom Length in Feet 36	40	50	*60	70
0°	17,600 (29.7)	13,600 (34.2)	7,750 (44.2)	4,010 (54.6)	2,670 (64.2)

Note: ( ) Reference radii in feet. A6-829-101047

#LMI operating code. Refer to LMI manual for instructions.
*60 ft. boom length is with inner-mid extended and outer-mid & fly retracted.

38302-11_F22.EPS

Figure 22 Capacity differences.

4.3.3 On Crawlers

Crawlers can be extended or retracted for lifting operations. Capacities do not exceed 75 percent of tipping on crawler cranes. Some cranes, however, may have outrigger extensions along with crawlers where capacities can be 85 percent of tipping. The operator should consult the operator's manual and follow the manufacturer's recommendations when lifting.

4.4.0 Tower and Ring Attachments

A tower attachment is used as an alternative to a freestanding tower crane where a lower erection height allows its use (*Figure 27*). It is also used to increase the load radius where reach is the primary operational concern. As with all attachments, the operator should follow the crane manufacturer's and/or attachment manufacturer's recommendations for lifting.

A ring attachment is used to greatly increase the crane's capacity through configuration changes and weight transfer to the ring foundation (*Figure 28*).

4.5.0 Counterweight Configurations

Each crane has counterweight specifications. Counterweights give extra stability to the crane during assembly and lifting operations. There

36-70 ft. | 12,150 lbs | Stationary | 360°

Pounds

#9005

Feet	36	40	50	*60	70
	Main Boom Length in Feet				
10	45,300 (69)	39,700 (72)			
12	41,750 (65.5)	39,700 (68.5)	29,600 (73.5)		
15	29,350 (59.5)	26,450 (63.5)	26,450 (70)	20,900 (74)	
20	17,800 (49)	17,650 (55)	17,050 (63.5)	16,250 (69)	16,250 (72)
25	11,750 (36)	11,700 (45)	11,350 (56.5)	10,850 (63.5)	10,850 (68)
30		8,040 (31.5)	7,820 (48.5)	7,470 (57.5)	7,470 (63)
35			5,400 (40)	5,120 (51.5)	5,120 (58)
40			3,660 (28)	3,200 (45)	3,430 (53)
45				1,000 (37)	2,150 (47.5)
50					1,150 (41.0)

Lifting Capacities at Zero Degree Boom Angle
On Rubber - Stationary 360

Boom Angle	36	40	50
	Main Boom Length in Feet		
0°	8,180 (29.7)	5,890 (34.2)	2,170 (44.2)

Note: ( ) Reference radii in feet. A6-829-101048A

#LMI operating code. Refer to LMI manual for instructions.
*60 ft. boom length is with inner-mid extended and outer-mid & fly retracted.

36-70 ft. | 12,150 lbs | Stationary | Defined Arc Over Front

Pounds

#9005

Feet	36	40	50	*60	70
	Main Boom Length in Feet				
10	45,300 (69)	42,850 (72)	29,600 (76)		
12	43,650 (65.5)	41,350 (68.5)	29,600 (73.5)		
15	38,300 (59.5)	36,300 (63.5)	29,600 (70)	20,900 (74)	17,300 (76.5)
20	31,150 (49)	29,550 (55)	25,900 (63.5)	20,900 (69)	17,300 (72)
25	24,100 (36)	24,150 (45)	21,800 (56.5)	18,800 (63.5)	17,300 (68)
30		17,400 (31.5)	17,200 (48.5)	15,300 (57.5)	15,300 (63)
35			12,800 (40)	12,500 (51.5)	11,000 (58)
40			9,720 (28)	9,390 (45)	9,390 (53)
45				7,090 (37)	7,090 (47.5)
50				5,310 (26.5)	5,310 (41)
55					3,870 (33.5)
60					3,170 (24)

Lifting Capacities at Zero Degree Boom Angle
On Rubber - Defined Arc Over Front

Boom Angle	36	40	50	*60	70
	Main Boom Length in Feet				
0°	17,600 (29.7)	13,600 (34.2)	7,750 (44.2)	4,010 (54.6)	2,670 (64.2)

Note: ( ) Reference radii in feet. A6-829-101047

#LMI operating code. Refer to LMI manual for instructions.
*60 ft. boom length is with inner-mid extended and outer-mid & fly retracted.

36-70 ft. | 12,150 lbs | Pick & Carry Up to 2.5 mph | Boom Center Over Front

Pounds

#9006

Feet	36	40	50	*60	70
	Main Boom Length in Feet				
10	42,500 (69)	37,600 (72)	30,700 (76)		
12	42,500 (65.5)	37,600 (68.5)	30,700 (73.5)	23,450 (77)	
15	37,600 (59.5)	37,600 (63.5)	30,700 (70)	19,700 (74)	16,250 (76.5)
20	32,950 (49)	32,900 (55)	30,700 (63.5)	19,700 (69)	16,250 (72)
25	24,100 (36)	24,150 (45)	23,850 (56.5)	19,700 (63.5)	16,250 (68)
30		17,400 (31.5)	17,200 (48.5)	16,800 (57.5)	16,250 (63)
35			12,800 (40)	12,500 (51.5)	12,500 (58)
40			9,720 (28)	9,390 (45)	9,390 (53)
45				7,090 (37)	7,090 (47.5)
50				5,310 (26.5)	5,310 (41)
55					3,870 (33.5)
60					3,090 (24)

Lifting Capacities at Zero Degree Boom Angle
On Rubber - Pick & Carry

Boom Angle	36	40	50	*60	70
	Main Boom Length in Feet				
0°	17,600 (29.7)	13,600 (34.2)	7,750 (44.2)	4,010 (54.6)	2,670 (64.2)

Note: ( ) Reference radii in feet. A6-829-101049

#LMI operating code. Refer to LMI manual for instructions.
*60 ft. boom length is with inner-mid extended and outer-mid & fly retracted.

NOTES:

1. Capacities are in pounds and do not exceed 75% of tipping loads as determined by test in accordance with SAE J765.
2. Capacities are applicable to machines equipped with 29.6x25 (28 ply) tires at 65 psi cold inflation pressure.
3. Defined Arc – Over front includes 6' on either side of longitudinal centerline of machine (ref. drawing C6-829-003529).
4. Capacities appearing above the bold line are based on structural strength and tipping should not be relied upon as a capacity limitation.
5. Capacities are applicable only with machine on firm level surface.
6. On rubber lifting with boom extensions not permitted.
7. For pick and carry operation, boom must be centered over front of machine, mechanical swing lock engaged and load restrained from swinging. When handling loads in the structural range with capacities close to maximum ratings, travel should be reduced to creep speeds.
8. Axle lockouts must be functioning when lifting on rubber.
9. All lifting depends on proper tire inflation, capacity and condition. Capacities must be reduced for lower tire inflation pressures. See lifting capacity chart for tire used. Damaged tires are hazardous to safe operation of crane.
10. Creep – Not over 200 ft. of movement in any 30 minute period and not exceeding 1 mph.

38302-11_F23.EPS

Figure 23 On-rubber load/capacity chart.

40 - 125 ft. (12 - 38 m) | 11,500 lbs. (5216 kg) | 0% 10' 4" Spread | 360°

Pounds

Feet	40	45	55	65	75	85	95	105	115	125
10	91,900 (70)	86,850 (72.5)								
12	70,100 (67)	66,600 (70)	61,200 (74)							
15	50,500 (61.5)	48,150 (65.5)	44,550 (70.5)	41,550 (74)						
20	31,200 (52.5)	30,950 (58)	28,900 (65)	26,950 (69)	26,600 (72.5)	26,600 (75)				
25	20,850 (41.5)	20,550 (49.5)	19,900 (58.5)	18,500 (64.5)	18,500 (68.5)	18,800 (71.5)	18,950 (73.5)	19,050 (75.5)		
30	14,650 (26)	14,350 (39.5)	13,850 (52)	12,950 (59)	13,850 (64)	13,650 (67.5)	13,950 (70.5)	14,200 (72.5)	14,200 (75)	*14,200 (76.5)
35		10,200 (26)	9,750 (44.5)	9,090 (53.5)	9,370 (59.5)	9,970 (64)	10,350 (67)	10,700 (69.5)	10,750 (72)	10,850 (74)
40			6,810 (35.5)	6,190 (47.5)	6,550 (54.5)	7,200 (60)	7,660 (63.5)	8,070 (66.5)	8,200 (69.5)	8,350 (71.5)
45			4,600 (23)	3,950 (40.5)	4,350 (49.5)	5,040 (55.5)	5,540 (60)	5,990 (63.5)	6,170 (66.5)	6,360 (69)
50				2,170 (32.5)	2,610 (44)	3,320 (51.5)	3,850 (56.5)	4,330 (60.5)	4,540 (64)	4,750 (66.5)
60							1,300 (48.5)	1,820 (53.5)	2,070 (58)	2,320 (61.5)
Minimum boom angle (deg.) for indicated length (no load)										51
Maximum boom length (ft.) at 0 degree boom angle (no load)										65

NOTE: ( ) Boom angles are in degrees.
*Based on maximum obtainable boom angle.

A6-829-011927A

Boom Angle	40	45	55
0°	12,550 (32.3)	8,440 (37.8)	3,610 (47.8)

NOTE: ( ) Reference radii are in feet.

A6-829-012282

38302-11_F24.EPS

Figure 24 Capacities for on outrigger at no extension (0 percent).

can be front, rear, side, or upper counterweights. Manufacturers specify in the operational notes or with specific load/capacity charts that the capacities listed are with counterweights in place. The counterweight sections must be combined with each other only in the way specified in the load charts. If a counterweight is extendable, it must be extended according to the manufacturer's recommendations. *Figures 29* through *31* show examples of counterweight configurations on different models of cranes.

4.6.0 Deduction Charts

In addition to the actual load weight, allowance must be made for additional weight. For example, the weight of the load block, headache ball, wire rope, parts of hoist line, and any boom extensions may require deduction from the gross capacity of the crane. These are all items that the crane will interpret as load. They are listed in charts supplied by the manufacturer (*Figure 32*).

40 - 125 ft.
(12 - 38 m)

11,500 lbs.
(5216 kg)

50%
17' 4" Spread

360°

Pounds

Feet	40	45	55	65	75	85	95	105	115	125
10	116,000 (70)	105,000 (72.5)								
12	102,500 (67)	102,000 (70)	94,600 (74)							
15	86,800 (61.5)	86,350 (65.5)	85,800 (70.5)	71,050 (74)						
20	67,250 (52.5)	64,850 (58)	60,550 (65)	56,950 (69)	55,250 (72.5)	48,150 (75)				
25	45,750 (41.5)	44,950 (49.5)	42,050 (58.5)	39,700 (64.5)	39,100 (68.5)	38,900 (71.5)	38,000 (73.5)	33,350 (75.5)		
30	32,500 (26)	32,050 (39.5)	31,000 (52)	29,250 (59)	29,050 (64)	29,250 (67.5)	29,300 (70.5)	29,250 (72.5)	24,550 (75)	*23,700 (76.5)
35		23,900 (26)	23,350 (44.5)	22,250 (53.5)	22,300 (59.5)	22,700 (64)	22,900 (67)	23,050 (69.5)	21,700 (72)	21,900 (74)
40			17,650 (35.5)	17,200 (47.5)	17,400 (54.5)	17,900 (60)	18,250 (63.5)	18,550 (66.5)	18,550 (69.5)	18,550 (71.5)
45			13,400 (23)	13,000 (40.5)	13,600 (49.5)	14,300 (55.5)	14,700 (60)	15,050 (63.5)	15,150 (66.5)	15,250 (69)
50				9,850 (32.5)	10,450 (44)	11,450 (51.5)	11,900 (56.5)	12,350 (60.5)	12,450 (64)	12,600 (66.5)
60					5,970 (30)	6,940 (41.5)	7,840 (48.5)	8,310 (53.5)	8,500 (58)	8,700 (61.5)
70						4,020 (28.5)	4,840 (39)	5,410 (46)	5,700 (51.5)	5,930 (55.5)
80							2,630 (27)	3,170 (37)	3,620 (44.5)	3,870 (49.5)
90								1,490 (25.5)	2,000 (36)	2,280 (42.5)
100										1,020 (34.5)
Minimum boom angle (deg.) for indicated length (no load)										27.5
Maximum boom length (ft.) at 0 degree boom angle (no load)										105

NOTE: ( ) Boom angles are in degrees.
*Based on maximum obtainable boom angle.

A6-829-011802A

Boom Angle	40	45	55	65	75	85	95
0°	22,800 (32.3)	18,250 (37.8)	11,500 (47.8)	6,220 (57.8)	3,740 (67.8)	2,350 (77.8)	1,300 (87.8)

NOTE: ( ) Reference radii are in feet.

A6-829-012282

38302-11_F25.EPS

Figure 25 Capacities for on outrigger at intermediate extension (50 percent).

4.6.1 Jib Charts

Charts provided by a manufacturer, such as the one shown in *Figure 33*, show the capacities in pounds of the luffing- or fixed-jib boom extensions at different degrees of offset. The capacity of a crane equipped with a jib boom extension is affected by the following:

- Load radius
- Length of the main boom
- Main boom angle
- Jib length
- Jib angle to the horizontal
- Jib angle with respect to the boom
- Weight of the rigging and hooks on both the jib and main boom

40 - 125 ft. (12 - 38 m)	11,500 lbs. (5216 kg)	100%	360°

Pounds

Feet	40	45	55	65	75	85	95	105	115	125
10	130,000 (70)	105,000 (72.5)								
12	111,000 (67)	105,000 (70)	94,600 (74)							
15	91,450 (61.5)	91,000 (65.5)	88,250 (70.5)	71,050 (74)						
20	69,550 (52.5)	69,050 (58)	68,400 (65)	60,400 (69)	55,250 (72.5)	48,150 (75)				
25	55,050 (41.5)	54,600 (49.5)	53,950 (58.5)	53,450 (64.5)	47,950 (68.5)	41,700 (71.5)	38,000 (73.5)	33,350 (75.5)		
30	42,950 (26)	42,450 (39.5)	41,700 (52)	41,200 (59)	41,950 (64)	36,700 (67.5)	33,300 (70.5)	30,750 (72.5)	24,550 (75)	*23,700 (76.5)
35		33,700 (26)	33,000 (44.5)	32,500 (53.5)	33,250 (59.5)	32,600 (64)	29,550 (67)	27,300 (69.5)	21,700 (72)	21,900 (74)
40			26,650 (35.5)	26,150 (47.5)	26,900 (54.5)	27,850 (60)	26,450 (63.5)	24,450 (66.5)	19,350 (69.5)	20,300 (71.5)
45			21,750 (23)	21,300 (40.5)	22,050 (49.5)	23,000 (55.5)	23,700 (60)	22,000 (63.5)	17,450 (66.5)	18,800 (69)
50				17,500 (32.5)	18,250 (44)	19,150 (51.5)	19,900 (56.5)	19,850 (60.5)	15,800 (64)	17,050 (66.5)
60					12,400 (30)	13,250 (41.5)	14,100 (48.5)	14,650 (53.5)	13,250 (58)	14,150 (61.5)
70						9,190 (28.5)	9,910 (39)	10,400 (46)	10,850 (51.5)	11,350 (55.5)
80							6,930 (27)	6,740 (37)	7,850 (44.5)	8,290 (49.5)
90								5,170 (25.5)	5,600 (36)	6,010 (42.5)
100									3,880 (25)	4,250 (34.5)
110										2,840 (24)
Minimum boom angle (deg.) for indicated length (no load)										0
Maximum boom length (ft.) at 0 degree boom angle (no load)										125

NOTE: ( ) Boom angles are in degrees.
*Based on maximum obtainable boom angle.

A6-829-011801A

Boom Angle	40	45	55	65	75	85	95	105	115
0°	22,800 (32.3)	18,250 (37.8)	12,200 (47.8)	7,990 (57.8)	5,720 (67.8)	4,320 (77.8)	3,210 (87.8)	2,380 (97.8)	1,570 (107.8)

NOTE: ( ) Reference radii are in feet.

A6-829-012282

38302-11_F26.EPS

Figure 26 Capacities for on outrigger at full extension (100 percent).

- Mathematical determination of whether the capacity of the crane is based on these factors:
  - Strength of the jib
  - Strength of the main boom
  - Stability of the crane

When a jib and/or boom extension is stowed on the heel section of the main boom, the stowed weight must be deducted from the gross capacity listed on the load/capacity chart in order to calculate the crane's net capacity.

When the jib/extension is mounted on the boom tip, the effective weight must be deducted from the gross capacity. This deduction is made only when the hoisting is done from the main boom. If you are lifting from an erected jib, do not count the weight of the jib as part of the load. It has already been compensated for in the jib's specific capacity chart. Effective weight is calculated by the manufacturer as the weight which, when applied at the boom tip, will have the same effect on the crane as the boom extension and/or jib.

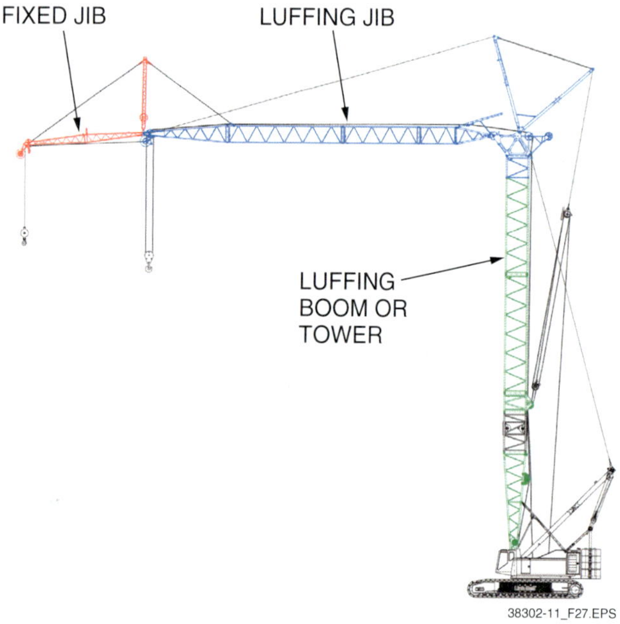

Figure 27 Crane with tower attachment.

Figure 29 Crawler crane counterweights.

Figure 28 200-ton ringer crane.

Figure 30 Ringer crane counterweights.

Figure 31 Counterweights on an auxiliary carrier.

CAPACITY DEDUCTIONS FOR AUXILIARY LOAD HANDLING EQUIPMENT

Load Handling Equipment:	(lbs.)
Auxiliary Head Attached	100
25-ton quick reeve 3 sheave hook block (see hook block for actual weight)	670
40-ton quick reeve 4 sheave hook block (see hook block for actual weight)	780
8.5-ton hook ball (see hook ball for actual weight)	360
Lifting From Main Boom With:	**(lbs.)**
28.5 ft. or 51 ft. fly stowed on base (see operation note 4)	0
28.5 ft. offset fly erected but not used	2600
51 ft. offset fly erected but not used	4800
Lifting From 28.5 ft. Offset Fly With:	
22.5 ft. fly tip erected but not used	PROHIBITED
22.5 ft. fly tip stowed on 28.5 ft. offset fly	PROHIBITED
Note: Capacity deductions are for Link-Belt supplied equipment only.	

38302-11_F32.EPS

Figure 32 Deductions chart.

4.6.2 Parts of Line

Before any lift is made, the operator must ensure that the crane has sufficient net capacity to lift the load and that it is also reeved with enough parts of line (*Figure 34*) to lift the load without breaking the hoist rope.

There are several methods used to determine the parts of line for a lift. Some crane manufacturers provide this information along with the load/capacity charts (*Figure 35*). This is the easiest of the methods. This method is also accurate if all the material used is recommended by the manufacturer.

On some crane charts, the working load limit (WLL) of the hoist rope may be reduced for multiple-part reeving due to block- and boom-point sheave friction. On other cranes, the determination of the parts of line may be in direct proportion to the number of layers on the winch drum. The greater the number of layers on the drum, the less line pull available to lift the load.

You can use the following formula to calculate the parts of line:

$$\text{Parts of line} = \frac{\text{suspended weight}}{\text{WLL of hoist rope}}$$

Where:

Suspended weight = load weight + weight of hook block + weight of slings and rigging hardware

$$\text{WLL} = \text{diameter squared} \times 8$$

Example:

- Size of wire rope: 1¼ inches
- Load weight: 50 tons
- Type of sheaves: anti-friction bearing

$$\text{WLL} = \text{diameter squared} \times 8 = 1.25 \times 1.25 \times 8 = 12.5$$

Total weight of load = 50

$$50 \text{ tons} \div 12.5 = 4 \text{ (ratio)}$$

Looking at *Table 1*, find the column under Ratio B, anti-friction bearing sheaves. Note that 4.71 is the next highest number over the calculated number of 4. The number of line parts column gives 5 parts of line.

5.0.0 Calculating Crane Capacity

Providing information on particular lifting configurations and presenting load/capacity charts applicable to those configurations is useless unless the math required to find capacities can be performed. For this reason, this section will demonstrate, step by step, how to find the gross and net capacities for a crane.

Crane manufacturers provide operational notes (notes for lifting capacities or other information) with the load charts for each crane model. Refer to *Figure 33* for an example. These notes stress the important facts of load/capacity charts and lifting covered in this module. It is important that the operator read and understand the notes because they contain stipulations regarding the operation of the crane.

Table 1 Ratio Factors

Ratio A Bronze Bushed Sheaves	Ratio B Anti-Friction Bearing Sheaves	Number Of Line Parts
.96	.98	1
1.87	1.94	2
2.75	2.88	3
3.59	3.81	4
4.39	4.71	5
5.16	5.60	6
5.90	6.47	7
6.60	7.32	8
7.27	8.16	9
7.91	8.98	10
8.52	9.79	11
9.11	10.60	12
9.68	11.40	13
10.20	12.10	14
10.70	12.90	15
11.20	13.60	16
11.70	14.30	17
12.20	15.00	18
12.60	15.70	19
13.00	16.40	20

38302-11_T01.EPS

33-56 ft. luffing folding boom extension (mode B) (fixed offset angles)

37.3-141.7 ft.	33 - 56 ft.	22,000 lbs	100% 34'-6" Spread	360°

Pounds

Feet	33 ft. LENGTH 5° OFFSET #0091	33 ft. LENGTH 20° OFFSET #0091	33 ft. LENGTH 40° OFFSET #0091	56 ft. LENGTH 5° OFFSET #0092	56 ft. LENGTH 20° OFFSET #0092	56 ft. LENGTH 40° OFFSET #0092
40	*13,700 (78)					
45	13,700 (77)					
50	13,700 (75)	13,700 (77.5)		*8,200 (78)		
55	13,700 (73.5)	13,700 (75.5)	*11,000 (78)	8,200 (77.5)		
60	13,700 (71.5)	13,700 (74)	11,000 (76)	8,200 (76)		
65	13,700 (70)	12,850 (72)	10,950 (74.5)	8,200 (74.5)	8,200 (77.5)	
70	12,500 (68)	12,000 (70)	10,350 (72.5)	8,200 (73)	8,200 (76)	
75	11,350 (66)	11,200 (68)	9,830 (70.5)	8,200 (71.5)	8,100 (74)	6,400 (77.5)
80	9,730 (64.5)	10,450 (66.5)	9,330 (68.5)	8,200 (69.5)	7,600 (72.5)	6,400 (76)
85	8,300 (62.5)	8,980 (64.5)	8,860 (66.5)	8,200 (68)	7,150 (71)	6,230 (74)
90	7,060 (60.5)	7,660 (62.5)	8,210 (64.5)	7,740 (66.5)	6,730 (69)	5,920 (72.5)
95	5,960 (58.5)	6,500 (60.5)	6,980 (62)	7,130 (64.5)	6,350 (67.5)	5,640 (70.5)
100	4,990 (56.5)	5,470 (58)	5,880 (60)	6,130 (63)	6,000 (65.5)	5,380 (68.5)
105	4,120 (54)	4,560 (56)	4,900 (58)	5,230 (61)	5,690 (64)	5,140 (67)
110	3,340 (52)	3,730 (54)	4,020 (55.5)	4,430 (59.5)	5,290 (62)	4,900 (65)
115	2,640 (49.5)	2,990 (51.5)	3,230 (53)	3,700 (57.5)	4,490 (60)	4,690 (63)
120	2,000 (47.5)	2,320 (49)	2,510 (50.5)	3,040 (55.5)	3,760 (58.5)	4,470 (61)
125	1,420 (45)	1,700 (46.5)	1,850 (47.5)	2,440 (53.5)	3,100 (56.5)	3,710 (58.5)
130		1,140 (44)	1,250 (45)	1,900 (51.5)	2,500 (54.5)	3,030 (56.5)
135				1,390 (49.5)	1,940 (52)	2,390 (54)
140					1,420 (50)	1,810 (52)
145						1,270 (49)
Minimum boom angle (°) for indicated length (no load)	42	43	43	48	48	47
Maximum boom length (ft.) at 0° boom angle (no load)		89.8			76.7	

NOTE: ( ) Boom angles are in degrees.

A6-829-103522

#LMI operating code. Refer to LMI manual for operating instructions.
*This capacity is based upon maximum boom angle.

NOTES:

1. All capacities above the bold line are based on structural strength of boom extension and do not exceed 85% of tipping loads, in accordance with SAE J-765.
2. The 33 ft. luffing folding boom extension may be used for single or double line lifting service. The 56 ft. luffing folding boom extension may be used for single line lifting service only.
WARNING: Lifting with the 33 ft. extension base, with the 23 ft. extension fly either erected or folded along side of extension base, is strictly prohibited.
3. WARNING: Operation of this machine with heavier loads than the capacities listed is strictly prohibited. Machine tipping with boom extension occurs rapidly and without advance warning.
4. Boom angle is the angle above or below horizontal of the longitudinal axis of the boom base section after lifting rated load.
5. Capacities listed are with outriggers properly extended and vertical jacks set only.
6. For main boom lengths less than 141.7 ft. with the boom extension erected, the rated loads are determined by boom angle. Use only the column which corresponds to the boom extension length and offset for which the machine is set up. For boom angles not shown, use rating of the next lower boom angle.
7. When lifting over the main boom nose with 33 ft. or 56 ft. extension erected, the outriggers must be fully extended or 50% extended (17.3 ft. spread).

38302-11_F33.EPS

Figure 33 Offsetable jib/boom extension chart.

Example 1:

When lifting from the main boom, find the net capacity of a Grove TM-1500 truck-mounted hydraulic crane (*Figure 36*) under the following conditions:

- Outriggers properly set and level
- Over-the-rear lift
- Power-pinned fly retracted
- 33- to 58-ft extension stowed
- Boom length of 52 ft
- Load radius of 23 ft
- 150-ton, 8-sheave block
- Auxiliary boom head attached
- 10-ton headache ball reeved over auxiliary boom head
- Rigging and other accessories of 195 pounds

The net capacity is 123,984 lbs. How was the net capacity determined? Refer to the load/capacity charts in *Figure 37* and the weight reduction charts in *Figure 38*.

> **WARNING!**
>
> Operators must follow load/capacity chart information according to manufacturer's specifications.

Read down the Radius column to 23 ft (*Figure 37*). This number is not listed, so go to the next longer load radius, which is 25 ft. Read across the Length row to 52 ft. It is not listed, so go to the boom length with the next lower gross capacity, which is 58 ft. The gross capacity where these two intersect is 131,500 lbs. Look up the deduction weights in the deduction charts (*Figure 38*), and subtract these from the gross capacity.

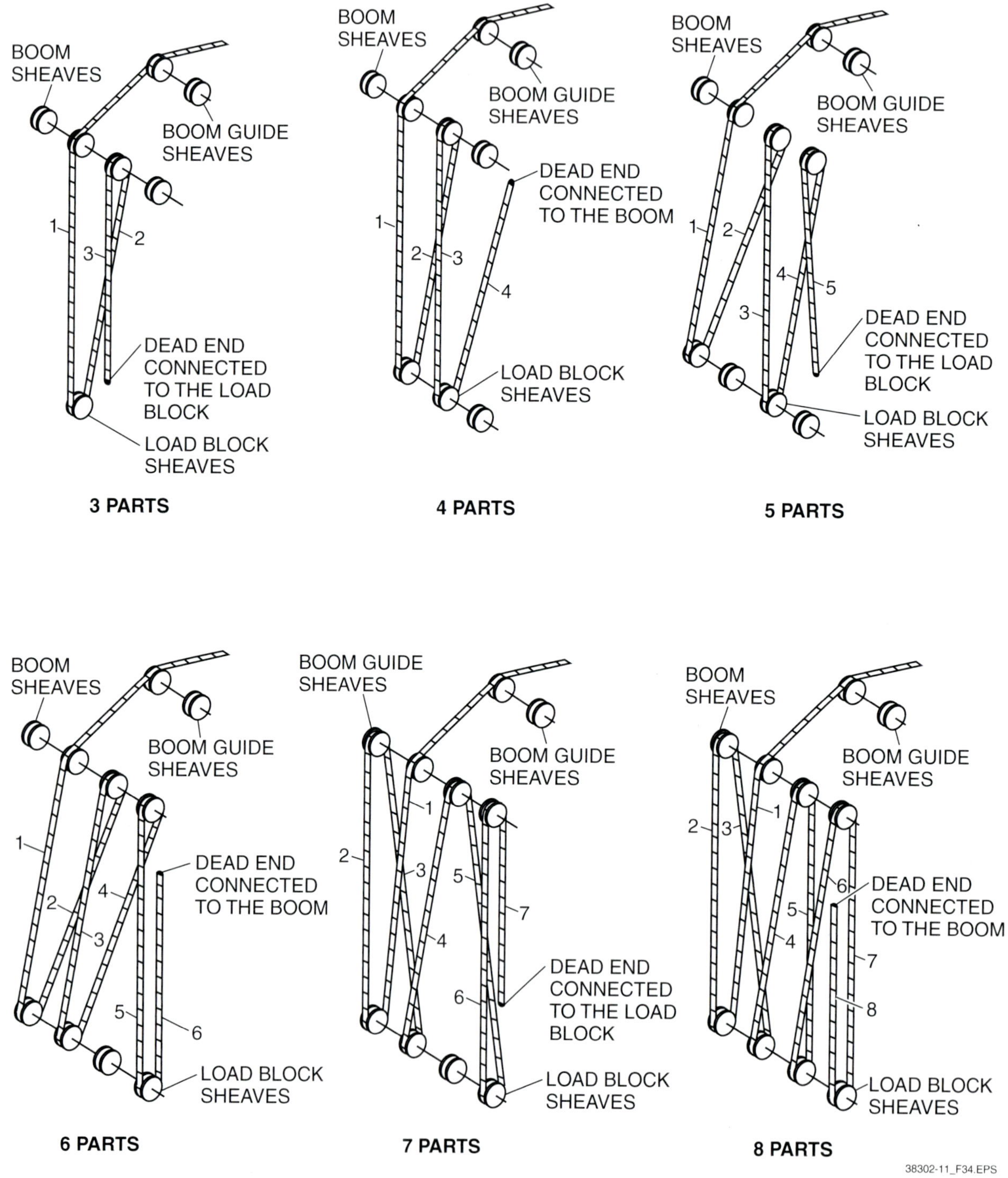

Figure 34 Reeving for various parts of line.

131,500 lbs (gross capacity)
– 1,246 lbs (extension stowed)
– 5,254 lbs (block)
– 261 lbs (aux. boom head)
– 560 lbs (headache ball)
– 195 lbs (rigging)

123,984 lbs (net capacity)

Example 2:

Find the net capacity of the Manitowoc 4100W-2 in the following configuration:

- Crane counterweights of 146,400 lbs installed
- 22C boom, 230-ft main boom no offset
- 125 ft of load radius
- Two-sheave boom point attached
- 15-ton hook block 20 ft below boom point
- 100-ton block reeved with 4 parts of line
- Crawlers extended with two 12,000-lb side counterweights installed

WIRE ROPE CAPACITY

Maximum Lifting Capacities Based On Wire Rope Strength			
Parts of Line	5/8" Type RB	5/8" Type ZB	Notes
1	9,080	11,080	Capacities shown are in pounds and working loads must not exceed the ratings on the capacity charts in the Crane Rating Manual.
2	18,160	22,160	
3	27,240	33,240	
4	36,320	44,320	
5	45,400	55,400	
6	54,480	66,480	Study Operator's Manual for wire rope inspection procedures and single part of line applications.
7	63,560	77,560	
8	72,640	88,640	
9	81,720		

LBCE	DESCRIPTION
TYPE RB	18 × 19 Rotation Resistant– Compact Strand, High Strength Preformed, Right Regular Lay
TYPE ZB	36 × 7 Rotation Resistant– Extra Improved Plow Steel – Right Regular Lay

38302-11_F35.EPS

Figure 35 Wire rope capacity chart.

38302-11_F36.EPS

Figure 36 Grove TM-1500 crane.

- Rigging of 345 lbs
- Lifting from the main hoist
- Two 30,000-lb carbody counterweights installed

The net capacity is 17,382 lbs. How was the net capacity determined? Refer to the Manitowoc 4100W-2 load/capacity charts (*Figure 39*).

Read the section of the load/capacity chart that applies to 230 ft of boom length. Read down the Operating Radius column. The radius of 125 ft has a gross capacity rating of 24,100 lbs. Deductions for rope and equipment below the boom or attached to the boom must be made for a lattice boom crane per the notes or information furnished with the load charts.

Weight of rope:

Boom tip height (boom point elevation) = 202.6

Whip line = 20' × 2.34 lbs = 46.8 lbs

Load line = 202.6' × 2.34 lbs = 474.1 lbs × 4 lines = 1,896.4 lbs

Total line weight = 1,896.4 lbs + 46.8 lbs = 1,943.2 lbs (rounded to 1,943 lbs)

24,100 lbs (gross capacity)
– 1,500 lbs (two-sheave point)
– 865 lbs (15-ton hook block)
– 2,065 lbs (100-ton block)
– 345 lbs (rigging)
– 1,943 lbs (rope)

17,382 lbs (net capacity)

6.0.0 Critical Lifts

A designated person must classify each lift into one of the following categories (ordinary, critical, or pre-engineered production) prior to planning the lift. A lift must be designated as critical if collision, upset, or dropping could result in any one of the following:

- Unacceptable risk of personnel injury or significant adverse health impact (on site or off site)
- Significant release of radioactive or other hazardous material or other undesirable conditions
- Undetectable damage that would jeopardize future operations or the safety of a facility
- Damage that would result in unacceptable delay to schedule or other significant program impact, such as loss of vital data

A lift should also be designated as critical if the load requires exceptional care in handling because of size, weight, close-tolerance installation, high susceptibility to damage, or other unusual factors.

Rated Lifting Capacities in Pounds
46 Ft – 173 Ft Boom on Outriggers - Over Rear

Radius in Feet	#0001									#0002
	Main Boom Length in Feet (Power Pinned Fly Retracted)									Power Pin. Fly Ext. & 141 ft.
	46	58	70	82	94	106	118	130	141	173
10	300,000 (74.5)									
12	280,000 (72)	143,500 (76)	142,000 (79)							
15	235,000 (67.5)	143,500 (72.5)	141,500 (76.5)	130,000 (78.5)						
20	173,500 (60.5)	143,500 (67.5)	123,500 (72)	112,000 (75)	102,000 (77.5)	90,300 (79.5)				
25	136,500 (52)	131,500 (61.5)	110,500 (67.5)	98,650 (71)	89,250 (74)	78,550 (76.5)	73,700 (78.5)	69,300 (80)		
30	106,000 (43)	106,000 (55.5)	98,000 (63)	88,350 (67.5)	78,750 (71)	69,250 (73.5)	65,100 (76)	61,000 (77.5)	60,000 (79.5)	
35	84,700 (30.5)	84,700 (49)	84,700 (58)	80,150 (63.5)	69,000 (67.5)	60,750 (70.5)	57,150 (73)	54,000 (75.5)	52,150 (77.5)	
40		70,500 (41)	70,500 (52.5)	70,500 (59.5)	61,300 (64)	54,000 (67.5)	50,600 (70.5)	48,300 (73)	45,850 (75)	38,000 (79)
45		58,850 (32)	58,850 (47)	58,850 (55)	55,000 (60.5)	48,500 (64.5)	46,200 (68)	43,050 (71)	40,400 (73)	35,750 (77)
50		49,600 (17.5)	49,600 (40.5)	49,600 (50.5)	48,750 (57)	43,050 (61.5)	40,700 (65)	38,250 (68.5)	35,750 (71)	32,100 (75.5)
60			36,200 (22.5)	36,200 (39.5)	36,200 (48.5)	34,300 (55)	33,600 (59.5)	30,750 (63.5)	28,500 (66.5)	26,350 (72)
70				26,050 (25)	26,050 (39.5)	26,050 (47.5)	26,050 (53)	24,750 (58)	23,100 (61.5)	22,000 (68.5)
80					18,850 (27)	18,850 (39)	18,850 (46.5)	18,850 (52.5)	18,700 (56.5)	18,500 (64.5)
90						13,500 (28)	13,500 (38.5)	13,500 (46.5)	13,500 (51.5)	15,250 (60.5)
100							9,390 (29)	9,390 (38)	9,390 (45.5)	12,600 (56.5)
110							6,080 (12.5)	6,080 (30.5)	6,080 (39)	10,100 (52)
120								3,390 (17.5)	3,390 (31)	7,530 (47.5)
130									1,150 (19.5)	5,390 (42.5)
140										3,610 (36.5)
150										2,100 (30)
Minimum boom angle (deg.) for indicated length (no load)									10	19
Maximum boom length (ft.) at 0 deg. boom angle (no load)									140	167

Note: ( ) Boom angles are in degrees.
#LMI operating code. Refer to LMI manual for instructions.

38302-11_F37.EPS

Figure 37 TM-1500 load/capacity chart.

For critical lifts, the operating organization must appoint a rigging superintendent or other qualified person to be present at the lift site during the entire lifting operation. The rigging supervisor or other qualified person must ensure that a lift plan or procedure (see *Appendix A*) that defines the operation is prepared and includes the following:

- Identification of the items to be moved; the weight, dimensions, and center of gravity of the load, and any hazardous or toxic materials that are present
- Identification of operating equipment to be used by type and rated capacity
- Rigging sketches that include (as applicable):
  - Identification and rated capacities of slings, lifting bars, rigging accessories, and below-the-hook lifting devices
  - Load-indicating devices
  - Lifting points
  - Sling angles
  - Boom and swing angles
  - Methods of attachment
  - Crane orientations
  - Other factors affecting equipment capacity
- Operating procedures and special instructions to operators, including rigging precautions and safety measures to be followed

Experienced operators who have been trained and qualified to operate the specific equipment to be used must be assigned to make the lift.

Only designated, qualified signalers must give signals to the operator. However, the operator must obey an Emergency Stop signal at all times, no matter who gives the signal.

WEIGHT REDUCTIONS FOR LOAD HANDLING DEVICES

33 ft. Extension	
*Stowed -	892 lbs.
*Erected -	5,704 lbs.

33 ft. - 58 ft. Extension	
*Stowed -	1,246 lbs.
*Erected (Ret.) -	8,412 lbs.
*Erected (Ext.) -	11,406 lbs.

46 ft. - 173 ft. Boom with	
*46 ft. Jib Erected -	9,613 lbs.
*60 ft. Jib Erected -	14,571 lbs.
*74 ft. Jib Erected -	20,443 lbs.
*88 ft. Jib Erected -	27,199 lbs.
*Fixed Jib Accessories -	327 lbs.

*Reduction of main boom capacities

HOOKBLOCKS:	
30 Ton, 1 Sheave	1,022 lbs.
150 Ton, 8 Sheave	5,254 lbs.
Auxiliary Boom Head	261 lbs.
10 Ton Headache Ball	560 lbs.
15 Ton Headache Ball	803 lbs.

38302-11_F38.EPS

Figure 38 TM-1500 weight reduction charts.

The procedure and rigging sketches must be reviewed and approved by the responsible manager (or designee) and the responsible oversight organization (such as safety, quality assurance, or quality control) before the lift is made.

MANITOWOC 4100W-2

LIFTCRANE CAPACITIES

BOOM NO. 22C WITH OPEN THROAT TOP
146,400 LB. CRANE COUNTERWEIGHT
60,000 LB. CARBODY COUNTERWEIGHT
26'6" CRAWLERS EXTENDED

WARNING: This chart will apply only when two 12,000 lb. side ctwts. and two 30,000 lb. carbody ctwts. bear MEC registered Serial Numbers.

LIFTING CAPACITIES: Capacities for various boom lengths and operating radii may be based on percent of tipping, strength of structural components, operating speeds and other factors.

Capacities are for freely suspended loads and do not exceed 75% of a static tipping load. Capacities based on structural competence are shown by shaded areas.

Capacities are shown in pounds. Deduct 1200 pounds from capacities listed when single sheave upper boom point is attached and 1500 pounds when two sheave upper boom point is attached. To comply with B30.5 requirements, upper boompoint cannot be used on the 260 ft. boom. Weight of jib, (see chart A), all load blocks, hooks, weight ball, slings, hoist lines beneath boom and jib point sheaves, etc., is considered part of the main boom load. Boom is not to be lowered beyond radii where combined weights are greater than rated capacity. Where no capacity is shown, operation is not intended or approved.

OPERATING CONDITIONS: Machine to operate in a level position on a firm surface with crawlers fully extended and gantry in working position and be rigged in accordance with and under conditions referred to in rigging drawing No. 190693 and load line specification chard No. 6592-A.

Crane operator judgement must be used to allow for dynamic load effects of swinging, hoisting or lowering, travel, as well as adverse operating conditions & physical machine depreciation.

OPERATOR RADIUS: Operating is the horizontal distance from the axis of rotation to the center of vertical hoist line or load block with the load freely suspended. Add 14" to boom point radius for radius of sheave when using single part hoist line.

Boom angle is the angle between horizontal and centerline of boom butt and inserts and is an indication of operating radius. In all cases, operating radius shall govern capacity.

BOOM POINT ELEVATION: Boom point elevation, in feet, is the vertical distance from ground level to centerline of boom point shaft.

MACHINE EQUIPMENT: Machine equipped with 26'6" extendible crawlers, 48" treads, 17' retractable gantry, 12 part boom hoist reeving, four 1 3/8" boom pendants, 1st ctwt. 41,900 lbs., 2nd ctwt. 41,500 lbs., 3rd ctwt. 39,000 lbs., two 12,000 lbs. side ctwt's. and two 30,000 lbs. carbody ctwt's.

LOAD AND WHIP LINE SPECIFICATIONS	
LOAD LINE:	1-1/8" - 6 x 31 Warrington-Seale, Extra Improved Plow Steel, Regular Lay, IWRC. Minimum Breaking Strength 65 Ton. (Approx.Weight Per Ft. in Lbs. 2.34)
WHIP LINE:	1-1/8" - Warrington-Seale, Improved Plow Steel, Regular Lay, IWRC. Minimum Breaking Strength 56.5 Ton. Maximum Load - 28,300 Lbs. Per Line. (Approx. Weight Per Ft. in Lbs. 2.34)

MAXIMUM BOOM AND JIB LENGTH LIFTED UNASSISTED				DEDUCT FROM CAPACITIES WHEN JIB IS ATTACHED	
OVER FRONT OF BLOCKED CRAWLERS		OVER SIDE OF EXTENDED CRAWLERS			
BOOM LENGTH	JIB NO. 123	BOOM LENGTH	JIB NO. 123	JIB LENGTH	JIB NO. 123
260'	--	260'	--	30'	3,000 lbs.
250'	--	250'	--	40'	3,600 lbs.
240'	40'	240'	40'	50'	4,200 lbs.
230'	60'	230'	60'	60'	4,900 lbs.

Load block, hook and weight ball on ground to start.

HOIST REEVING FOR MAIN LOAD BLOCK						
No.Parts Of Line	1	2	3	4	5	6
Max Load - Lbs.	32,500	65,000	97,500	130,000	162,500	195,000
No. Parts of Line	7	8	9	10	11	12
Max. Load - Lbs.	227,500	260,000	292,500	325,000	357,500	400,000
No. Parts of Line	13					
Max. Load - Lbs.	430,000					

FOR JIB CAPACITIES, CONSULT JIB CHART.

BOOM LGTH FEET	OPER. RAD. FEET	BOOM ANG. DEG.	BOOM POINT ELEV.	CAPACITY: CRAWLERS EXTENDED
70	16.5	79.7	75.9	460,000
	17	79.3	75.8	400,000
	18	78.5	75.6	380,100
	19	77.6	75.4	363,000
	20	76.8	75.1	347,300
	22	75.1	74.6	319,600
	24	73.4	74.1	293,400
	26	71.7	73.5	266,100
	28	69.9	72.8	237,500
	30	68.2	72.0	214,300
	32	66.4	71.2	195,100
	34	64.6	70.2	178,900
	36	62.8	69.3	165,200
	38	60.9	68.2	153,300
	40	59.1	67.0	143,000
	45	54.1	63.7	122,100
	50	48.9	59.8	106,300
	55	43.2	54.9	93,900
	60	36.9	49.0	84,000
	65	29.4	41.3	75,800
	70	19.5	30.3	63,900

BOOM LGTH FEET	OPER. RAD. FEET	BOOM ANG. DEG.	BOOM POINT ELEV.	CAPACITY: CRAWLERS EXTENDED
80	17	80.6	85.9	392,800
	18	79.9	85.8	378,900
	19	79.2	85.6	361,800
	20	78.5	85.4	346,100
	22	77.0	84.9	318,400
	24	75.5	84.5	292,500
	26	74.0	83.9	265,600
	28	72.5	83.3	237,000
	30	71.0	82.7	213,800
	32	69.5	81.9	194,600
	34	68.0	81.2	178,500
	36	66.4	80.3	164,700
	38	64.8	79.4	152,800
	40	63.3	78.4	142,400
	45	59.2	75.7	121,500
	50	54.9	72.5	105,700
	55	50.4	68.6	93,300
	60	45.6	64.1	83,400
	65	40.3	58.8	75,200
	70	34.4	52.2	68,300
	75	27.4	43.9	52,500
	80	18.2	32.0	53,900

BOOM LGTH FEET	OPER. RAD. FEET	BOOM ANG. DEG.	BOOM POINT ELEV.	CAPACITY: CRAWLERS EXTENDED
90	18	81.1	95.9	355,400
	19	80.4	95.7	346,900
	20	79.8	95.6	336,900
	22	78.5	95.2	317,400
	24	77.2	94.7	291,700
	26	75.9	94.3	264,800
	28	74.5	93.7	236,600
	30	73.2	93.2	213,400
	32	71.9	92.5	194,200
	34	70.5	91.9	178,000
	36	69.2	91.1	164,200
	38	67.8	90.3	152,300
	40	66.4	89.5	142,000
	45	62.9	87.1	121,100
	50	59.3	84.4	105,200
	55	55.5	81.2	92,800
	60	51.5	77.5	82,900
	65	47.3	73.2	74,700
	70	42.8	68.2	67,800
	75	37.9	62.3	62,000
	80	32.4	55.2	57,000
	85	25.8	46.2	52,600
	90	17.1	33.5	45,900

BOOM LGTH FEET	OPER. RAD. FEET	BOOM ANG. DEG.	BOOM POINT ELEV.	CAPACITY: CRAWLERS EXTENDED
100	19	81.4	105.9	332,900
	20	80.8	105.7	327,100
	22	79.6	105.4	316,200
	24	78.5	105.0	290,800
	26	77.3	104.5	263,900
	28	76.1	104.1	236,200
	30	74.9	103.6	212,900
	32	73.7	103.0	193,700
	34	72.5	102.4	177,500
	36	71.3	101.7	163,700
	38	70.1	101.0	151,800
	40	68.9	100.3	141,400
	45	65.8	98.2	120,500
	50	62.6	95.8	104,700
	55	59.3	93.0	92,300
	60	55.9	89.8	82,300
	65	52.4	86.2	74,100
	70	48.7	82.1	67,200
	75	44.8	77.4	61,400
	80	40.5	72.0	56,400
	85	35.9	65.6	52,000
	90	30.7	58.0	48,200
	95	24.5	48.5	44,900
	100	16.3	35.0	39,300

CAUTION! CHECK AMOUNT OF COUNTERWEIGHT ON MACHINE BEFORE USE OF THIS CHART

38302-11_F39A.EPS

Figure 39 Manitowoc 4100 W-2 load/capacity chart (1 of 4).

LIFTCRANE CAPACITIES – 4100W SERIES 2 BOOM NO 22C WITH OPEN-THROAT TOP, CONTINUED

BOOM LGTH FEET	OPER. RAD. FEET	BOOM ANG. DEG.	BOOM POINT ELEV.	CAPACITY: CRAWLERS EXTENDED
110	22	82.3	115.7	240,500
	24	81.3	115.4	232,100
	26	80.2	115.0	224,400
	28	79.1	114.6	217,300
	30	78.1	114.1	210,700
	32	77.0	113.6	193,300
	34	75.9	113.1	177,100
	36	74.8	112.5	163,300
	38	73.7	111.8	151,400
	40	72.6	111.2	141,000
	45	69.9	109.3	120,100
	50	67.0	107.2	104,200
	55	64.1	104.7	91,800
	60	61.2	101.9	81,800
	65	58.1	98.8	73,600
	70	54.9	95.3	66,800
	75	51.6	91.3	60,900
	80	48.1	86.8	55,900
	85	44.4	81.7	51,600
	90	40.4	75.9	47,800
	95	36.1	69.2	44,400
	100	31.1	61.1	41,400
	105	25.3	51.1	38,700
	110	17.6	37.2	33,900
120	22	83.0	125.9	228,700
	24	82.0	125.5	220,500
	26	81.0	125.2	213,000
	28	80.0	124.8	206,100
	30	79.1	124.4	199,700
	32	78.1	123.9	192,800
	34	77.1	123.4	176,600
	36	76.1	122.9	162,800
	38	75.1	122.3	150,900
	40	74.1	121.7	140,500
	45	71.6	120.0	119,500
	50	69.1	118.1	103,700
	55	66.5	115.9	91,300
	60	63.8	113.4	81,300
	65	61.1	110.6	73,100
	70	58.3	107.5	66,200
	75	55.4	104.0	60,400
	80	52.4	100.1	55,300
	85	49.2	95.8	51,000
	90	45.9	91.0	47,200
	95	42.4	85.5	43,800
	100	38.6	79.4	40,800
	105	34.4	72.2	38,100
	110	29.7	63.7	35,700
	115	24.1	53.2	33,500
130	24	82.6	135.7	212,700
	26	81.7	135.3	205,300
	28	80.8	135.0	198,600
	30	79.9	134.6	192,300
	32	79.0	134.2	186,500
	34	78.1	133.7	176,300
	36	77.2	133.2	162,500
	38	76.3	132.7	150,600
	40	75.4	132.1	140,200
	45	73.1	130.6	119,300
	50	70.8	128.8	103,400
	55	68.4	126.8	91,000
	60	66.0	124.6	81,000
	65	63.5	122.0	72,800
	70	61.0	119.2	65,900
	75	58.4	116.1	60,000
	80	55.7	112.7	55,000
	85	53.0	108.9	50,700
	90	50.1	104.8	46,800
	95	47.1	100.1	43,500
	100	43.9	95.0	40,500
	105	40.6	89.2	37,800
	110	36.9	82.6	35,300
	115	33.0	75.1	33,100
	120	28.5	66.1	31,100
	125	23.1	55.1	29,300
140	26	82.3	145.5	198,400
	28	81.5	145.1	191,800
	30	80.7	144.8	185,600
	32	79.8	144.4	179,900
	34	79.0	144.0	174,700
	36	78.1	143.5	162,000
	38	77.3	143.0	150,100
	40	76.5	142.5	139,700
	45	74.3	141.1	118,700
	50	72.2	139.4	102,800
	55	70.0	137.6	90,400
	60	67.8	135.5	80,400
	65	65.6	133.2	72,200
	70	63.3	130.7	65,300
	75	60.9	127.9	59,500
	80	58.5	124.8	54,400
	85	56.1	121.4	50,100
	90	53.5	117.7	46,200
	95	50.9	113.7	42,900
	100	48.1	109.2	39,900
140	105	45.3	104.2	37,200
	110	42.2	98.8	34,700
	115	39.0	92.7	32,500
	120	35.5	85.7	30,500
	125	31.7	77.8	28,700
	130	27.4	68.5	27,000
	135	22.2	57.0	25,500
150	28	82.1	155.3	185,600
	30	81.3	154.9	179,600
	32	80.5	154.6	174,000
	34	79.7	154.2	168,800
	36	79.0	153.8	161,600
	38	78.2	153.3	149,600
	40	77.4	152.8	139,200
	45	75.4	151.5	118,200
	50	73.4	150.0	102,400
	55	71.4	148.3	89,900
	60	69.4	146.4	79,900
	65	67.3	144.3	71,700
	70	65.2	141.9	64,800
	75	63.0	139.4	59,000
	80	60.9	136.6	53,900
	85	58.6	133.5	49,600
	90	56.3	130.1	45,700
	95	54.0	126.5	42,400
	100	51.5	122.5	39,400
	105	49.0	118.2	36,700
	110	46.4	113.4	34,200
	115	43.6	108.2	32,000
	120	40.7	102.4	30,000
	125	37.6	96.0	28,200
	130	34.2	88.8	26,500
	135	30.5	80.5	25,000
	140	26.4	70.8	23,500
	145	21.4	58.8	22,200
160	28	82.6	165.4	179,800
	30	81.8	165.1	173,900
	32	81.1	164.7	168,400
	34	80.4	164.4	163,300
	36	79.7	164.0	158,600
	38	78.9	163.6	149,100
	40	78.2	163.1	138,700
	45	76.3	161.9	117,700
	50	74.5	160.5	101,800
	55	72.6	158.9	89,300
	60	70.7	157.1	79,300
	65	68.8	155.1	71,100
	70	66.8	153.0	64,200
	75	64.9	150.6	58,400
	80	62.9	148.0	53,300
	85	60.8	145.2	49,000
	90	58.7	142.2	45,100
	95	56.6	138.8	41,800
	100	54.4	135.2	38,700
	105	52.1	131.4	36,000
	110	49.7	127.1	33,600
	115	47.3	122.5	31,400
	120	44.8	117.5	29,400
	125	42.1	112.0	27,600
	130	39.3	105.9	25,900
	135	36.3	99.2	24,300
	140	33.1	91.7	22,900
	145	29.5	83.0	21,600
	150	25.5	72.9	20,400
	155	20.7	60.5	19,200
170	30	82.3	175.2	169,000
	32	81.6	174.9	163,600
	34	81.0	174.5	158,700
	36	80.3	174.2	154,000
	38	79.6	173.8	148,800
	40	78.9	173.3	138,400
	45	77.2	172.2	117,400
	50	75.4	170.9	101,500
	55	73.7	169.4	89,100
	60	71.9	167.7	79,000
	65	70.1	165.9	70,800
	70	68.3	163.9	63,900
	75	66.4	161.7	58,100
	80	64.6	159.3	53,000
	85	62.7	156.7	48,700
	90	60.8	153.9	44,900
	95	58.8	150.8	41,500
	100	56.8	147.5	38,500
	105	54.7	144.0	35,800
	110	52.6	140.2	33,300
	115	50.4	136.0	31,100
	120	48.1	131.5	29,100
	125	45.8	126.7	27,300
	130	43.3	121.4	25,600
	135	40.8	115.7	24,100
	140	38.1	109.3	22,600
	145	35.2	102.3	21,300
	150	32.0	94.5	20,100
	155	28.6	85.5	18,900
	160	24.7	75.1	17,800
	165	20.0	62.2	16,400
180	32	82.1	185.0	158,700
	34	81.5	184.7	154,000
	36	80.8	184.3	149,500
	38	80.2	184.0	145,200
	40	79.5	183.6	137,900
	45	77.9	182.5	116,900
	50	76.3	181.2	101,000
	55	74.6	179.8	88,500
	60	72.9	178.3	78,500
	65	71.3	176.6	70,300
	70	69.6	174.7	63,400
	75	67.8	172.6	57,500
	80	66.1	170.4	52,500
	85	64.3	168.0	48,100
	90	62.5	165.4	44,300
	95	60.7	162.5	40,900
	100	58.9	159.5	37,900
	105	57.0	156.2	35,200
	110	55.0	152.7	32,700
	115	53.0	149.0	30,500
	120	51.0	144.9	28,500
	125	48.9	140.5	26,700
	130	46.7	135.8	25,000
	135	44.4	130.7	23,500
	140	42.0	125.2	22,000
	145	39.6	119.2	20,700
	150	36.9	112.6	19,500
	155	34.1	105.3	18,300
	160	31.1	97.2	17,200
	165	27.7	87.9	16,200
	170	23.9	77.1	15,300
	175	19.4	63.8	13,500
190	32	82.5	195.1	153,100
	34	81.9	194.8	148,800
	36	81.3	194.5	144,800
	38	80.7	194.1	140,800
	40	80.1	193.8	137,000
	45	78.5	192.7	116,400
	50	77.0	191.5	100,500
	55	75.4	190.2	88,000
	60	73.9	188.8	78,000
	65	72.3	187.1	69,700
	70	70.7	185.4	62,800
	75	69.1	183.4	57,000
	80	67.4	181.4	52,000
	85	65.8	179.1	47,600
	90	64.1	176.6	43,800
	95	62.4	174.0	40,400
	100	60.7	171.2	37,400
	105	58.9	168.2	34,700
	110	57.1	164.9	32,200
	115	55.3	161.4	30,000
	120	53.4	157.7	28,000
	125	51.5	153.7	26,200
	130	49.5	149.5	24,500
	135	47.5	144.9	22,900
	140	45.3	140.0	21,500
	145	43.1	134.6	20,200
	150	40.8	128.9	18,900
	155	38.4	122.6	17,800
	160	35.9	115.8	16,700
	165	33.2	108.3	15,700
	170	30.2	99.8	14,800
	175	27.0	90.3	13,900
	180	23.3	79.1	13,100
	185	18.9	65.4	11,000
200	34	82.3	204.9	143,100
	36	81.7	204.6	139,500
	38	81.2	204.3	136,000
	40	80.6	203.9	132,500
	45	79.1	203.0	115,900
	50	77.7	201.8	100,000
	55	76.2	200.6	87,500
	60	74.7	199.2	77,400
	65	73.2	197.7	69,200
	70	71.7	196.0	62,300
	75	70.2	194.2	56,400
	80	68.6	192.2	51,400
	85	67.1	190.1	47,000
	90	65.5	187.8	43,200
	95	63.9	185.3	39,800
	100	62.3	182.7	36,800
	105	60.6	179.8	34,100
	110	59.0	176.8	31,600
	115	57.3	173.6	29,400
	120	55.5	170.2	27,400
	125	53.0	166.5	25,600
	130	51.9	162.6	23,900
	135	50.1	158.4	22,300
	140	48.1	153.9	20,900
	145	46.2	149.1	19,600
	150	44.1	144.0	18,300
	155	42.0	138.4	17,200
	160	39.7	132.5	16,100
	165	37.4	126.0	15,100
	170	34.9	118.9	14,200
200	175	32.3	111.1	13,300
	180	29.4	102.4	12,400
	185	26.2	92.5	11,700
	190	22.6	81.0	10,900
	195	18.4	67.0	8,500
210	36	82.1	214.7	134,200
	38	81.6	214.4	131,000
	40	81.0	214.1	127,900
	45	79.6	213.2	115,500
	50	78.3	212.1	99,600
	55	76.9	210.9	87,100
	60	75.4	209.6	77,100
	65	74.0	208.1	68,800
	70	72.6	206.6	61,900
	75	71.2	204.8	56,100
	80	69.7	203.0	51,000
	85	68.2	201.0	46,700
	90	66.7	198.8	42,800
	95	65.2	196.5	39,400
	100	63.7	194.0	36,400
	105	62.2	191.3	33,700
	110	60.6	188.5	31,300
	115	59.0	185.5	29,100
	120	57.4	182.3	27,100
	125	55.7	178.9	25,200
	130	54.1	175.2	23,500
	135	52.3	171.4	22,000
	140	50.6	167.3	20,600
	145	48.8	162.9	19,200
	150	46.9	158.2	18,000
	155	45.0	153.2	16,800
	160	43.0	147.9	15,800
	165	40.9	142.1	14,800
	170	38.7	135.9	13,800
	175	36.5	129.2	12,900
	180	34.0	121.9	12,100
	185	31.5	113.9	11,300
	190	28.7	104.9	10,600
	195	25.6	94.7	9,900
	200	22.1	82.9	8,800
	205	17.9	68.5	6,500
220	38	82.0	224.6	125,700
	40	81.4	224.2	122,800
	45	80.1	223.3	115,000
	50	78.8	222.3	99,100
	55	77.5	221.2	86,600
	60	76.1	220.0	76,600
	65	74.8	218.6	68,300
	70	73.4	217.1	61,400
	75	72.0	215.4	55,500
	80	70.7	213.7	50,500
	85	69.3	211.8	46,100
	90	67.9	209.7	42,300
	95	66.4	207.5	38,900
	100	65.0	205.2	35,800
	105	63.6	202.7	33,100
	110	62.1	200.0	30,700
	115	60.6	197.2	28,500
	120	59.1	194.2	26,500
	125	57.5	191.0	24,600
	130	55.9	187.6	23,000
	135	54.3	184.0	21,400
	140	52.7	180.2	20,000
	145	51.0	176.1	18,600
	150	49.3	171.8	17,400
	155	47.6	167.3	16,300
	160	45.7	162.4	15,200
	165	43.9	157.2	14,200
	170	41.9	151.7	13,200
	175	39.9	145.7	12,300
	180	37.8	139.3	11,500
	185	35.6	132.4	10,700
	190	33.2	124.8	10,000
	195	30.7	116.6	9,300
	200	28.0	107.3	8,600
	205	25.0	96.9	8,000
	210	21.5	84.8	6,500

CAPACITIES CONTINUED ON NEXT PAGE

These load charts are intended for instructional purposes only. They were derived from manufacturer sales information which may not be complete or machine specific. Not responsible for typographical errors.

38302-11_F39B.EPS

Figure 39 Manitowoc 4100 W-2 load/capacity chart (2 of 4).

LIFTCRANE CAPACITIES – 4100W SERIES 2 BOOM NO 22C WITH OPEN-THROAT TOP, CONTINUED

SEE CONDITIONS ON FRONT PAGE

BOOM LGTH FEET	OPER. RAD. FEET	BOOM ANG. DEG.	BOOM POINT ELEV.	CAPACITY: CRAWLERS EXTENDED
230	38	81.5	234.5	**132,900**
	40	81.0	234.2	**125,200**
	45	79.7	233.3	114,500
	50	78.5	232.4	98,600
	55	77.2	231.3	86,100
	60	75.9	230.1	76,100
	65	74.6	228.8	67,800
	70	73.3	227.3	60,900
	75	72.0	225.8	55,000
	80	70.7	224.1	50,000
	85	69.4	222.3	45,600
	90	68.0	220.3	41,700
	95	66.7	218.2	38,400
	100	65.3	216.0	35,300
	105	64.0	213.6	32,600
	110	62.6	211.1	30,200
	115	61.1	208.4	28,000
	120	59.7	205.6	26,000
	125	58.3	202.6	24,100
	130	56.8	199.4	22,400
	135	55.3	196.0	20,900
	140	53.8	192.5	19,400
	145	52.2	188.7	18,100
	150	50.6	184.7	16,900
	155	49.0	180.5	15,700
	160	47.3	176.0	14,700
	165	45.6	171.3	13,600
	170	43.8	166.2	12,700
	175	42.0	160.8	11,800
	180	40.1	155.1	11,000
	185	38.1	148.9	10,200
	190	36.0	142.3	9,400
	195	33.9	135.1	8,700
	200	31.6	127.3	**8,100**
	205	29.1	118.8	**7,400**
	210	26.4	109.3	**6,800**
240	40	81.4	244.3	**123,400**
	45	80.2	243.5	**112,600**
	50	78.9	242.6	98,100
	55	77.7	241.5	85,600
	60	76.5	240.4	75,600
	65	75.3	239.1	67,300
	70	74.0	237.7	60,400
	75	72.8	236.3	54,500
	80	71.5	234.6	49,500
	85	70.3	232.9	45,100
	90	69.0	231.1	41,200
	95	67.7	229.1	37,800
	100	66.4	227.0	34,800
	105	65.1	224.7	32,100
	110	63.8	222.3	29,700
	115	62.5	219.8	27,400
	120	61.1	217.1	25,400
	125	59.7	214.3	23,600
	130	58.3	211.3	21,900
	135	56.9	208.1	20,300
	140	55.5	204.7	18,900
	145	54.0	201.2	17,600
	150	52.5	197.5	16,300
	155	51.0	193.5	15,200
	160	49.5	189.4	14,100
	165	47.9	185.0	13,100
	170	46.2	180.3	12,200
	175	44.6	175.4	11,300
	180	42.8	170.2	10,400
	185	41.0	164.6	9,600
	190	39.2	158.7	8,900
	195	37.3	152.3	8,200
	200	35.2	145.5	**7,500**
	205	33.1	138.1	**6,700**
	210	30.9	130.1	**6,000**

BOOM LGTH FEET	OPER. RAD. FEET	BOOM ANG. DEG.	BOOM POINT ELEV.	CAPACITY: CRAWLERS EXTENDED
250	45	80.6	253.6	**106,600**
	50	79.4	252.7	97,800
	55	78.2	251.7	85,200
	60	77.1	250.6	75,200
	65	75.9	249.4	66,900
	70	74.7	248.1	60,000
	75	73.5	246.7	54,100
	80	72.3	245.2	49,100
	85	71.1	243.5	44,700
	90	69.9	241.7	40,800
	95	68.7	239.8	37,400
	100	67.4	237.8	34,400
	105	66.2	235.7	31,700
	110	64.9	233.4	29,300
	115	63.6	231.0	27,100
	120	62.4	228.5	25,000
	125	61.1	225.8	23,200
	130	59.7	222.9	21,500
	135	58.4	219.9	20,000
	140	57.0	216.8	18,500
	145	55.7	213.4	17,200
	150	54.3	209.9	16,000
	155	52.8	206.2	14,800
	160	51.4	202.4	13,700
	165	49.9	198.3	12,700
	170	48.4	193.9	11,800
	175	46.8	189.4	10,900
	180	45.3	184.5	10,100
	185	43.6	179.4	9,300
	190	41.9	174.0	8,500
	195	40.2	168.3	**7,800**
	200	38.4	162.2	**7,000**
	205	36.5	155.7	**6,200**
	210	34.5	148.6	**5,500**
260	45	80.9	263.7	**104,800**
	50	79.8	262.9	**95,700**
	55	78.7	261.9	84,700
	60	77.6	260.9	74,700
	65	76.4	259.7	66,400
	70	75.3	258.5	59,500
	75	74.2	257.1	53,600
	80	73.0	255.6	48,500
	85	71.8	254.1	44,100
	90	70.7	252.4	40,300
	95	69.5	250.6	36,900
	100	68.3	248.6	33,900
	105	67.1	246.6	31,100
	110	65.9	244.4	28,700
	115	64.7	242.1	26,500
	120	63.5	239.7	24,500
	125	62.3	237.1	22,600
	130	61.0	234.4	20,900
	135	59.7	231.6	19,400
	140	58.5	228.6	17,900
	145	57.2	225.4	16,600
	150	55.8	222.1	15,400
	155	54.5	218.7	14,200
	160	53.1	215.0	13,100
	165	51.7	211.2	12,100
	170	50.3	207.1	11,200
	175	48.9	202.9	10,300
	180	47.4	198.4	9,500
	185	45.9	193.7	8,700
	190	44.3	188.7	**7,900**
	195	42.7	183.4	**7,000**
	200	41.1	177.8	**6,200**
	205	39.4	171.9	**5,400**
	210	37.6	165.6	**4,700**

38302-11_F39C.EPS

Figure 39 Manitowoc 4100 W-2 load/capacity chart (3 of 4).

WEIGHTS 4100W Cont'd

DESCRIPTION	APPROX. WEIGHT (IN LBS.)
JIB NO. 123	
Jib Top - 15' (w/Jib point)	695
Jib Butt - 15'	690
Jib Insert - 10'	340
Basic Pendant - 33' 3-3/4" (2 Req'd)	115 each
Pendant - 10' (2 Per insert)	65 each
Jib Backstay Pendant	155 each
Jib Strut - 12'-6"	365
COMPONENTS	
Hook Rollers (6) - w/shafts	1,020
Light Plant - 6.5KW - w/mounting platform	1,390
Catwalk - left and right side w/rails	1,320
Lagging - 27-5/8" dia. plain	1,410
Boom Hoist Rope - 12 Part - 760' of 7/8" - 6 x 26	1,080
Wire Rope Guide Assembly - Lower	325
Wire Rope Guide Assembly - Upper	510
Rope Guide Roller Assembly	55 each
2-Part Gantry w/Telescopic Backhitch	7,805
Equalizer	2,000
Hoist Line - 1-1/8" - 6 x 31	2.34 lbs./ft.
Whip Line - 1-1/8" - 6 x 31	2.34 lbs./ft.
15-Ton Hook and Weight Ball	865
100-Ton Hook Block Assembly	2,065
200-Ton Hook Block Assembly	4,900
230-Ton Hook Block Assembly	5,375
Boom Stop - Telescopic Air Cushioned	675
Dragline Fairlead - Revolving	1,910
Dragline Fairlead - Hinged	9,330

NOTE: The above weights may fluctuate up or down 5% due to manufacturing tolerances.

38302-11_F39D.EPS

Figure 39 Manitowoc 4100 W-2 load/capacity chart (4 of 4).

7.0.0 Engineering Considerations

The following engineering requirements need to be considered when planning a lift:

- Basic load data, such as the object to be lifted, its dimensions and general configuration (rectangle, square, sphere, or irregular), and its gross weight
- Location in which load is to be lifted and placed
- Maximum operating radius
- Lift attachment locations and descriptions
- Maximum distance between any two attachment locations
- Sling lengths needed to keep the horizontal angle greater than 60 degrees
- Specifications of the crane to be used (manufacturer, model)
- Counterweight attachments

For lattice boom cranes:

- Counterweights used on the lower and upper works
- Number of boom sections used and their sizes and lengths
- Number of jib sections used and their sizes and lengths
- Maximum capacity at the maximum lift radius
- Hoist block capacity
- Wire rope size and weight per foot
- Rated single line pull
- Parts of line reeved to the block
- The maximum capacity of the lift point for the boom and the jib

Ground support:

- Ground pressure allowed at the setup location and the ground pressure created by the crane and load at all radii out to the planned maximum load radius
- The maximum allowable crane load that will not exceed the ground support

Load calculations:

- Weight of the load, crane components below the lifting points, and wire rope

Administration:

- Pre-lift meeting
- Completion of lift plan

8.0.0 Lift Plan Implementation

For a lift plan to be successful, the critical data must be communicated to all personnel involved. Holding a pre-lift meeting accomplishes this goal. The signal person(s), riggers, operators, management, and site supervisors should all be involved in this meeting. Officially identify which operator and signal person(s) will be assigned the job, and determine how communication is to be maintained during the lift (including the use of hand signals, radio, or both, or other methods). Also, determine whether tag lines will be used. Discuss all potential hazards, including weather, power lines, or mechanical failure. Review emergency procedures. Finally, have the person responsible for the lift, the site superintendent, the site safety manager, and any others needed sign off on the lift plan.

If required by the critical lift procedure, a test lift must be done before the critical lift. Conditions for a test lift should closely simulate actual conditions, which may include weight, rigging selection and configuration, and other relevant factors.

Summary

Lift plans are just as important to the success of a lifting operation as proper load/capacity chart interpretation. The operator, supervisor, signal person, and all other personnel involved in the lift must understand what is required on a lift plan and how it affects them. Personnel must also understand mathematical calculations and the equipment's limitations.

Review Questions

1. A document that outlines all weight, rigging, and configuration requirements for a lift is a _____.
   a. lift plan
   b. manufacturer's worksheet
   c. pre-lift qualification
   d. load chart

2. Lift plans are mandatory for _____.
   a. steel erection or multiple-crane lifts
   b. lifts using extensions or jibs
   c. luffing or derricking operations
   d. lifts using a load block

3. The number of shackles required is a criterion that falls under the category of _____.
   a. load weight
   b. sizing of slings
   c. wire ropes
   d. rope reeving

4. Foundation and location of electrical hazards are examples of _____ criteria.
   a. crane placement
   b. hoisting
   c. hydraulic
   d. pneumatic

5. The size and weight per foot of the _____ need to be considered as part of a lift plan.
   a. boom
   b. jib
   c. wire rope
   d. swing-away

6. Over-the-front, over-the-rear, and over-the-side are _____ considerations for crane placement.
   a. crawler configuration
   b. area of operation
   c. gantry position
   d. counterweight

7. Critical lifts always require _____.
   a. an equalizer beam
   b. beam clamps
   c. radio communication
   d. a lift plan

8. When lifting from the main boom of a lattice crane, an erected jib, whip line length below the jib, and the length of the line from the lift point to load block are calculations involving the _____.
   a. weight of crane components below the lifting point
   b. load block height
   c. rigging on the block
   d. effective weight of crane components past the lifting point

9. The load block weight is considered part of the weight of the _____.
   a. load to be lifted, not including rigging
   b. crane components below the lifting point
   c. crane components past the lifting point
   d. boom components

10. The effective weight of crane components past a main boom lifting point includes _____.
    a. an erected jib
    b. any parts of line considered part of the load
    c. the maximum wire rope weight to be experienced during the lift
    d. lifting beams, slings, and shackles

11. If the *net* capacity of a crane in a particular configuration is 95,000 lbs and the total weight of the deductions is 2,500 lbs, the *gross* capacity is _____.
    a. 92,500 lbs
    b. 95,000 lbs
    c. 97,500 lbs
    d. impossible to calculate without more information

12. If the *gross* capacity of a crane at 200 ft of boom length and 50 ft of load radius is 83,000 lbs, and the deductions are a 1,500-lb boom point, a 865-lb hook block, 335 lbs of rigging, and 1,900 lbs of rope, the *net* capacity is _____.
    a. 74,600 lbs
    b. 78,400 lbs
    c. 87,600 lbs
    d. impossible to calculate without more information

13. If the *gross* capacity of a crane is 90,000 lbs and its *net* capacity in a particular configuration is 83,425 lbs, the total weight of the deductions is _____.
    a. 2,875 lbs
    b. 4,932 lbs
    c. 6,575 lbs
    d. impossible to calculate without more information

14. By adjusting the crane and/or load attachment points on an equalizer beam, it is possible to _____.
    a. safely load a crane beyond its net capacity
    b. ensure that load rate exceeds the lowest rated net capacities of the cranes
    c. share the load between cranes in any proportion desired
    d. share the load between cranes equally

15. In a multiple-crane lift, the crane with the greatest net capacity must always be _____.
    a. equipped with single part hoist line
    b. loaded to at least 75 percent of its rated capacity
    c. farthest from the load's center of gravity
    d. closest to the load's center of gravity

16. When an equalizer beam is used, the load radius of each crane is measured to _____.
    a. the center of gravity of the load
    b. its pick-up point on the equalizer
    c. the center of the equalizer
    d. the attachment point of the load on the equalizer

17. In a multiple-crane lift using an equalizer, the weight of the equalizer is considered part of the load weight.
    a. True
    b. False

18. In a multiple-crane lift, if the total load weight is 100,000 lbs, and the total *net* capacity of Crane One is 80,000 lbs, the total *net* capacity of Crane Two must be no less than _____ lbs to meet the 75-percent capacity guidelines.
    a. 53,333
    b. 75,050
    c. 100,500
    d. 180,000

Refer to *Figure 1* for Questions 19 and 20.

19. If A = 20 ft, then what is the value for B?
    a. 30 ft
    b. 35 ft
    c. 40 ft
    d. 45 ft

20. If A= 10 ft and B = 15 ft, then what is the actual load on Crane One?
    a. 57,000 lbs
    b. 60,000 lbs
    c. 90,000 lbs
    d. 150,000 lbs

21. In preparing for multiple-crane lifting, the load blocks of each crane have to be positioned relative to the load's center of gravity using the formula _____.
    a. (net capacity Crane One) × A = (net capacity Crane Two) × B
    b. (net capacity Crane One) + A = (net capacity Crane Two) + B
    c. (net capacity Crane Two) × A = (net capacity Crane One) × B
    d. (net capacity Crane One) – A = (net capacity Crane Two) – B

22. To determine the share of the load weight on Crane One that is positioned a distance of (A) feet from the load's center of gravity in a multiple-crane lift, the operator must use the formula _____.
    a. [B ÷ (A + B)] × load weight
    b. [B ÷ (A – B)] × load weight
    c. [A ÷ (A + B)] × load weight
    d. [A ÷ (A – B)] × load weight

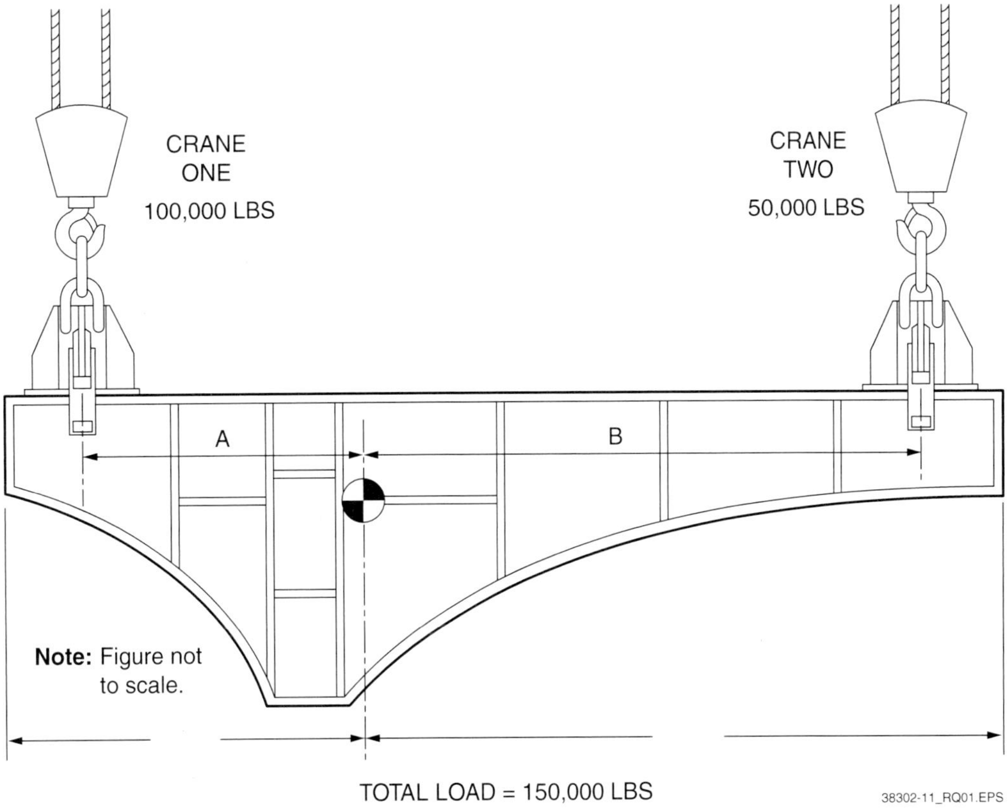

Figure 1

23. If the boom length falls between values listed on the load chart, the operator should use the capacity rating for the _____.
    a. smallest load shown at either the next larger radius or next longer boom length
    b. next shorter boom length
    c. next longer load radius to compensate for the boom length
    d. next shorter boom angle to compensate for the boom length

24. The boom point elevation is the distance from _____.
    a. the ground to the center of the boom point sheaves
    b. a projection of the axis of rotation to the center of the vertical hoist line
    c. the boom foot pins to the center of the boom point sheaves
    d. the bottom of the base section to the center line of the boom

25. On some hydraulic telescoping cranes, which different boom configurations are activated and deactivated by means of a switch?
    a. Boom modes
    b. Boom ranges
    c. Electrical extensions
    d. Mechanical extensions

26. Interpolation is permitted in determining load capacities.
    a. True
    b. False

27. As a crane swings to various quadrants, the distance that the load is positioned from the tipping axis _____.
    a. doesn't change
    b. results in no stability change
    c. may change dramatically
    d. can be corrected with proper interpolation

28. Cranes are typically built to be _____.

a. as wide as they are long
b. wider than they are long
c. longer than they are wide
d. wider in the back than in the front

29. Crane manufacturers produce a separate load/capacity chart for each operational quadrant.

a. True
b. False

30. Crane capacities must be reduced for lower tire inflation pressure.

a. True
b. False

31. For on-rubber lifting, capacities do *not* exceed _____ of tipping.

a. 70 percent
b. 75 percent
c. 85 percent
d. 95 percent

32. The _____ gives extra stability to the crane during lifting operations.

a. counterweight
b. auxiliary boom sheave
c. hoist rope
d. hook block

33. The weights of the load block, headache ball, and wire rope of a crane are _____.

a. weights that need to be added to the gross capacity
b. deductions that need to be subtracted from the gross capacity
c. already compensated for in the capacity chart
d. deducted from gross capacity only when hoisting is done from the main boom

34. Jib charts show the capacities of the jib boom extension _____.

a. with rigging and load attachments
b. at different degrees of offset
c. in the stowed position
d. with rigging attached

35. When lifting from the main boom with the jib extension mounted on the boom tip, the effective jib weight must be _____ the *gross* capacity.

a. added to
b. multiplied by
c. deducted from
d. divided by

36. To find the parts of line, the operator _____.

a. adds the suspended weight and the WLL of the hoist rope
b. divides the suspended weight by the WLL of the hoist rope
c. subtracts the WLL of the hoist rope from the suspended weight
d. divides the load weight by the WLL of the hoist rope

37. Suspended weight includes all of the following *except* _____.

a. hook block weight
b. sling and rigging hardware
c. hoist rope weight
d. load weight

38. Referring to the Grove RT522 load/capacity charts in *Appendix B*, find the *net* capacity of the Grove RT522 configured as follows:

- On outriggers, over-the-front lift
- 23- to 38-ft telescoping jib stowed on boom
- 40-ft main boom length
- 23-ft load radius
- 15-ton, 2-sheave block on main boom
- Auxiliary boom head attached
- 5-ton headache ball on auxiliary boom head
- Rigging and accessories of 65 lbs
- Lifting from the main boom

a. 16,522 lbs
b. 16,731 lbs
c. 16,918 lbs
d. 17,141 lbs

39. Referring to the Grove RT522 load/capacity charts in *Appendix B*, find the *net* load that can be handled safely with the RT522 configured as follows:

- On rubber, 360 degrees, stationary
- 20.5 × 25 tires
- 23-foot A-frame jib stowed on boom
- 22-foot load radius
- 12-ton, 1-sheave (12 ½" OD) block on main boom
- Rigging and accessories of 70 pounds

a. 3,211 lbs
b. 3,594 lbs
c. 3,728 lbs
d. 3,914 lbs

40. Referring to the Grove RT522 load/capacity charts in *Appendix B*, find the *net* capacity of the Grove RT522 when configured as follows:

- On outriggers fully extended, 360 degrees
- 46-ft boom length
- 15-ft load radius
- 22-ton, 3-sheave hook block
- Rigging and accessories of 45 pounds

a. 29,755 lbs
b. 29,555 lbs
c. 29,315 lbs
d. 29,050 lbs

41. Referring to the Manitowoc 4100W-2 load/capacity chart in *Appendix C*, find the *gross* capacity of the Manitowoc 4100W-2 in the following configuration:

- Crane counterweights of 146,400 lbs installed
- 22C open-throat top boom
- 180 ft of main boom
- Boom angle of 60 degrees
- Crawlers extended with two 12,000-lb side counterweights installed
- Two 30,000-lb carbody counterweights installed

a. 37,900 lbs
b. 40,900 lbs
c. 45,300 lbs
d. 52,100 lbs

42. Referring to the Manitowoc 4100W-2 load/capacity charts in *Appendix C*, find the *net* capacity in the following configuration:

- Crane counterweights of 146,000 lbs installed
- 22C boom, 190 ft of main boom, no offset
- 130 ft of load radius
- 2-sheave boom point attached
- 15-ton hook block 20 ft below boom point
- 100-ton block reeved with 4 parts of line
- Crawlers extended with two 12,000-lb side counterweights installed
- Rigging of 345 lbs
- Lifting from the main hoist
- Two 30,000-lb carbody counterweights installed

a. 17,213 lbs
b. 17,729 lbs
c. 17,811 lbs
d. 18,292 lbs

43. Prior to planning a lift, a designated person must classify the lift into which of the following categories?

a. Major, minor, critical
b. Ordinary, critical, pre-engineered production
c. Standard, critical
d. General, special, critical

44. During critical lift operation, the operator must obey a(n) _____ signal at all times.

a. Travel
b. Emergency Stop
c. Dog Everything
d. Swing

45. A test lift is recommended before _____.

a. any lift
b. a lift involving a luffing jib boom
c. a lift involving a 100-ton hook block assembly
d. a critical lift

Trade Terms Introduced in This Module

Boom point elevation: Distance from the ground to the center of the boom point sheaves.

Engineering: Practical application of the knowledge of physics, leverage, and gravity as they relate to lifting operations.

Equalizer: Any of the various devices or appliances for equalizing stresses, strains, and pressures.

Interpolation: Method of estimating or finding an intermediate value or term in a sequence or between two known values.

Lift plan: A written plan that outlines all weight, rigging, and configuration considerations for a lift.

Appendix A

Critical Lift Plan

APPENDIX A – CRITICAL LIFT PLAN

INDEX

1. OBJECTIVE

2. DEFINITIONS

2.1 Critical Lift
2.2 Minor Lift
2.3 Rigging
2.4 Lift Crane
2.5 Tail Crane

3. PROCEDURE

3.1 Responsibility for Critical Lift
3.2 Responsibility for Minor Lift
3.3 Special Lifting Requirements
3.4 Training

APPENDICES

Appendix 1 - Planned Lifting Procedure Form

PREPARED BY: ______________________
Engineering Representative

APPROVED BY: ______________________
Safety Representative

AUTHORIZED BY: ______________________
Project / Site Manager

DEVELOPED BY CRANE INSPECTION & CERTIFICATION BUREAU - 2004

1

38302-11_A01.EPS

1 PURPOSE

1.1 Define responsibilities for personnel when crane lifts are made.

- To ensure the rigging and lift crane or cranes have adequate capacity for handling the weight of the device to be lifted.
- To ensure that the personnel responsible for the lift are qualified and competent personnel.
- To protect personnel and property.

2 DEFINITIONS

2.1 **Critical Lift** – As an accepted "General Industry Standard" The combined weight of the load, rigging and crane configuration exceeds 75% - 85% of the capacity of the crane and or there are complications as a result of load configuration, clearance, danger to personnel, power transmission equipment, critical service equipment or hazardous materials.

- The weight of the load is 25 tons or greater and exceeds 50' in length, height, and width. (Detailed Lift Plan required)
- Lifts involving more than one crane to handle a common load.

2.2 **Minor Lift** - The combined weight of the device, rigging and crane configuration is less than 80% of the crane capacity and there are no compilations as above.

- The lift weight is less than 25 tons and 50' in length.
- Lift is single crane.

2.3 **Rigging** – Any component part used to connect the device to be lifted by the crane; i.e., choker, sling, shackles, jib, spreader bars, snatch block, etc.

2.4 **Lift Crane** – The primary crane used to ultimately support the total weight of the device.

2.5 **Tail Crane** - The equipment used to tail or maneuver the device while the lift crane supports the device.

38302-11_A02.EPS

3 PROCEDURE

3.1 Responsibility for Critical Lifts

3.1.1 The Rigging Superintendent making the lift is responsible for planning and executing the lift. The superintendent shall complete the attached lifting procedure and acquire all approval signatures prior to the lift. The original "approved" procedure is to be submitted to the organization's Engineering Representative and Safety director for record and one copy retained by the Construction Rigging Superintendent. The Rigging Superintendent must include the organization's Engineering Representative and Safety Director in developing the lift plan to ensure full consideration of existing conditions and circumstances. Any lifts exceeding 85% must have Project / Site Manager 's signature.

The planning should include:

- Determining the maximum weight of the device to be lifted.
- Calculating the combined weight of the device, rigging accessories, load blocks and crane jibs.
- Calculating the sling and rigging hardware capacities with consideration to load angle factors and stresses on all slings, hardware and lift points.
- Determining the required boom angle and swing radius.
- Selecting the proper crane for the lift and the tail crane when required.
- Inspecting the lift site to determine the soil condition to decide whether matting and / or soil stabilization is needed.
- Determine the area to be restricted during lift.

38302-11_A03.EPS

3.1.2 After a careful review of the job site, the hoisting equipment and critical lift plan, the Rigging Supervisor will record the project data on the Planned Lifting Procedure Form. The crane operator will sign and date the completed form. In addition to his signature, the lifting procedure will be reviewed and signed by others as indicated on the lift form.

3.1.3 The crane operator has the authority and responsibility to ensure the lift is proper and in accordance with established local procedures, as well as OSHA and ANSI standards.

3.2 Responsibility for Minor Lifts

3.2.1 The Rigging Superintendent or competent person making the lift is responsible for completing the "Lifting Procedure" planning / executing the lift and submitting the lifting procedure to the Engineering Representative for record. Note: Repetitive minor lifts need only one "Lifting Procedure" completed per day of lifting activities; i.e., multiple pieces of structural steel and pipe installation. However, if the crane is repositioned, another "Lifting Procedure" should be completed before lifting is resumed.

- Should hoisting activities become routinely cycle duty procedures and located in remote locations the Engineering Representative or Project Superintendent may waive the completion of a lift form after review.

The Planning Should Include:

- Determining the maximum weight of the device to be lifted.
- Selecting the rigging based upon the maximum weight and calculate sling and hardware stresses.
- Calculating the combined weight of the device, rigging, load blocks and crane jibs.
- Determining the center of gravity of the load.

38302-11_A04.EPS

- Determining the required boom length, boom angle and swing radius at the pick point and the point the load is to be placed.
- Selecting the proper crane for the lift and the tail crane when required.
- Inspecting the lift site to determine the soil condition to decide whether matting and/or soil stabilization is needed.
- Identify the location of all utilities.
- Determine the area to be restricted to public access during lift.
- Determine the swing area.

3.2.2 After the careful review of the job site, hoisting equipment and the lift plan is completed, the Rigging Supervisor will record the project data on a Planned Lifting Procedure Form. The crane operator will sign and date the completed form. In addition to his signature, the Lifting Procedure will be reviewed and signed by the rigger.

3.3 Special Lifting Requirements

Whenever lifting devices which are smooth surfaced and lightweight; i.e., fiberglass or aluminum vessels, with canvas or nylon slings a strongback and / or bracing sling shall be used to prevent the lifting slings from rendering. Rendering (or slippage) of the main lifting slings can occur if the resistant force of friction between the sling and vessel wall is exceeded due to lift movement. Therefore, the strongback and / or bracing slings must be properly attached to prevent the main lifting slings from rendering in any position during the entire lift.

3.4 Training

3.4.1 Each person whose job requires them to operate cranes shall receive the following training:

- The recognition of the hazards associated with lifting equipment.

38302-11_A05.EPS

- The fundamentals of rigging practices and the understanding of load capacity charts as they pertain to crane configurations.
- The method and means of calculating load capacities.
- Attend a training course in mobile crane operations and rigging considerations.

3.4.2 Each person who is involved in rigging shall be required to participate in a rigging training program and tested to become competent in the selection of rigging when a lift is performed.

3.4.3 All trades personnel and supervision involved in lifting shall be instructed about this procedure and the requirements.

38302-11_A06.EPS

3.4.4 PLANNED LIFTING PROCEDURE FORM

1. DATE OF FORM PREPARATION:	
2. PROJECT NUMBER:	
3. PROJECT TITLE / LOCATION:	
4. PERMIT REQUIRED: Yes / No	
5. PREPARED BY:	
6. LOAD DESCRIPTION / PHYSICAL SIZE:	
• Equipment Name	
• I.D. / Tag Number	
• Physical Dimensions	
• Equipment Orientation V or H	
• Lift Lug Orientation Degrees	
• Hoisted load / Device Age	
• Physical Condition	
• NDE Required Y / N	
7. LIFTING AND TAILING ATTACHMENTS APPROVED: Y / N	
8. LOAD ELEVATION:	
9. LOAD WEIGHT:	

38302-11_A07.EPS

APPLICATION - LIFT CRANE

10. CRANE(S) TYPE:

11. CRANE CONFIGURATION:

12. PARTS OF WIRE ROPE: Wire Rope Breaking Strength

- Single Line Pull Wire Rope SWL

13. MAXIMUM RADIUS:

14. BOOM ANGLE:

15. BOOM CLEARANCE CHECK REQUIRED: Yes No

16. TAIL SWING CLEARANCE VERIFIED: Yes No

17. CHART CAPACITY:

18. CRANE RIGGING:	SIZE	CAPACITY	CONDITION
• Choker(s)	________	________	________
• Sling(s)	________	________	________
• Shackle(s)	________	________	________
• Spreader Bar(s)	________	________	________
• Other			

19. CRANE RIGGING COMPONENT WEIGHT:

- Jib Weight
- Ball Weight
- Load Block Weight
- Suspended Load Line Weight
- Shackle(s) Weight
- Choker(s) Weight
- Sling(s) Weight
- Spreader Bar(s) Weight
- Other Lifting Attachment Weights
- Total Rigging Weight

20. GROSS LOAD:

21. % OF CHART:USED

DEVELOPED BY CRANE INSPECTION & CERTIFICATION BUREAU - 2004 8

38302-11_A08.EPS

APPLICATION - TAIL CRANE

10. CRANE(S) TYPE:

11. CRANE CONFIGURATION:

12. PARTS OF WIRE ROPE: Wire Rope Breaking Strength

- Single Line Pull Wire Rope SWL

13. MAXIMUM RADIUS:

14. BOOM ANGLE:

15. BOOM CLEARANCE CHECK REQUIRED: Yes No

16. TAIL SWING CLEARANCE VERIFIED: Yes No

17. CHART CAPACITY

18. CRANE RIGGING	SIZE	CAPACITY	CONDITION
• Choker(s)	______	______	______
• Sling(s)	______	______	______
• Shackle(s)	______	______	______
• Spreader Bar(s)	______	______	______
• Other			

19. CRANE RIGGING COMPONENT WEIGHT:

- Jib Weight
- Ball Weight
- Load Block Weight
- Suspended Load Line Weight
- Shackle(s) Weight
- Choker(s) Weight
- Sling(s) Weight
- Spreader Bar(s) Weight
- Other Lifting Attachment Weights
- Total Rigging Weight

20. GROSS LOAD:

21. % OF CHART USED:

22. RIGGING SAFETY DESIGN FACTOR 5 to 1 Yes No

DEVELOPED BY CRANE INSPECTION & CERTIFICATION BUREAU - 2004

9

38302-11_A09.EPS

CRITICAL LIFTS **NOTES**

Engineering Certification Spreader Bar (Required)

- Proof Test Certification of Chokers 1 1/4" &

Greater (Required)

23. SOIL TYPE / CONDITION:

24. MATS / CRIBBING Yes No

25. RIGGING INSPECTED: Yes No

26. TAG LINE:

27. HAZARDS

28. WEATHER:

29. DATE: TIME:

30. LIFT MEETING CONDUCTED: Yes No

31. APPROVALS:

Rigging Supervisor: ______________________________

Crane Operator. ______________________________

Rigger / Signalman. ______________________________

Client Representative (If Required):

Safety Representative (If Required):

Engineering Representative (If Required)

Project Manager (If Required)

Communications: What form of communications will be used and by whom?

DEVELOPED BY CRANE INSPECTION & CERTIFICATION BUREAU - 2004 10

302A10.EPS

INSTRUCTIONS FOR FILLING OUT PLANNED LIFTING PROCEDURE FORM

1. **Date of form preparation**.
2. **PROJECT NUMBER:** Enter the project number that has been assigned to this work activity.
3. **PROJECT TITLE/LOCATION:** Enter the name of the project, plant site or construction site of the lift.
4. **PERMIT REQUIRED (y/n):** Indicate if a permit will be required. Certain plant safe work procedures require a permit when making certain lifts over in-service facilities or lifts made near power lines or guy lines or exceeding 75% of load crane capacity chart.
5. **PREPARED BY:** Enter name of person responsible for research and preparation of this form.
6. **LOAD DESCRIPTION / PHYSICAL SIZE:** Describe the load to be lifted including name of the equipment and item number and or tag number. Enter physical dimensions of load to be lifted. Identify equipment orientation with a "V" or "H", "V" for vertical and "H" for horizontal. Identify age of existing equipment / device to be lifted, physical condition and / or if NDE is required on item to be lifted prior to lift.
7. **LIFTING / TAILING ATTACHMENTS APPROVED (y/n):** Indicate if lifting attachments are approved by, the equipment engineering firm.
8. **LOAD ELEVATION:** Indicate the maximum and minimum elevation in feet that the load will travel Enter the maximum radius the load will achieve during the lift cycle of pick, swing & set.
9. **LOAD WEIGHT:** Determine the weight of the load utilizing certified engineer drawings, vendor data or operations manual. Only use load weight information, which has been documented and certified, or calculated by a Mechanical or Vessel P.E.
10. **CRANE(S) TYPE:** Enter the type, model and rated capacity of lift equipment to be used for lifting
11. **CRANE CONFIGURATION:** Indicate the length of the main boom and the length of jib (if equipped) that will be in the lift & tail crane(s) at time of lift.
12. **PARTS OF WIRE ROPE / SINGLE LINE PULL:** Enter the number of parts in the load handling line during the lift. Enter the allowable winch single line pull in lbs.

DEVELOPED BY CRANE INSPECTION & CERTIFICATION BUREAU - 2004

11

38302-11_A11.EPS

13. **MAXIMUM RADIUS:** Enter the maximum working radius the crane / load will achieve during the lift cycle of pick, swing & set.

14. **BOOM ANGLE:** Indicate the cranes' boom angle at the initial pick point / radius and final set point/radius.

15. **BOOM ANGLE:** Indicate the cranes' boom angle at the initial pick point / radius and final set point / radius.

16. **TAIL SWING CLEARANCE VERIFIED (y/n):** Indicate if the tail swing clearance has been verified. When it is required, the sketch / diagram verifying the tail swing clearance should be attached to the PCR Planned Crane Lifting Procedure Form.

17. **CHART CAPACITY:** Indicate the capacity of the crane(s) from the manufacturer's capacity chart with the parameters as indicated in items (8) thru (9).

18. **CRANE RIGGING:** Enter the size of chokers and shackles used for the lift and their physical condition.

19. **RIGGING COMPONENT WEIGHTS:** Enter the manufacturer's effective weight for the jib, load block, suspended load line, ball, suspended whip / load line, shackles, slings, spreader bars and any other lifting attachments suspended under the boom point.

20. **GROSS LOAD:** Indicate the total weight of the load by adding (9) and (19).

21. **% OF CAPACITY CHART USED:** Indicate the percentage of the crane's chart capacity to be utilized for the lift. If load / capacity percentage exceeds 75%, the lift should not be made without review of PCR Engineering Representative and Project / Site Manager. "Net load divided by net capacity = % Capacity Used "

22. **RIGGING SAFETY FACTOR 5 TO 1 (y/n):** Verify that the rigging equipment (shackles, chokers, etc.) used to perform the lift have a 5 to 1 Safety Factor.

    - Engineering Certification on Spreader Bars
    - Proof Test Certifications of Shackles and Slings

23. **SOIL CONDITION:** Indicate the type of soil in the area of the lift; loose, compact, or undisturbed earth. Consideration should be given to any adjacent excavation and its distance from the outriggers or crawler tracks.

24. **MATS REQUIRED:** Depending on the soil conditions determined above, indicate if any matting is required to properly protect any underground facilities and to secure the affected area under the

38302-11_A12.EPS

lifting equipment. Matting location, configuration and the number of mats should be identified on the rigging plans and diagrams.

25. **RIGGING REVIEWED (y/n):** Indicate if the rigging has been reviewed. This includes chokers, shackles, spreaders, load line, lifting lugs and any other lifting attachments required. Depending on the nature of the lift, it may be required that a lift plan and diagram should be developed and attached to the "Planned Crane Lifting Procedure Form." This may be required at the discretion of the Client Representative or the Rigging Supervisor.

26. **TAG LINE (y/n)** Indicate if a tag lines will be required.

27. **HAZARDS:** Indicate if there are any existing power lines or overhead obstructions in the area of the lift. In addition, indicate where there are any existing underground utilities or items buried in the crane set-up area. Identify the hazard (water, sewage, drainage, electrical, etc.) and its depth.

28. **WEATHER:** Indicate the weather conditions prior to and during the lift with emphasis on wind velocity and direction, rain probability, etc.

29. **DATE: TIME:** Indicate the date and approximate time of day that the lift will be made.

30. **LIFT MEETING CONDUCTED (y/n):** Indicate if a pre-lift meeting will be conducted with all involved parties. The client representative may request such a meeting.

31. **APPROVALS:** A minimum of four (4) signatures will be required on the lift form. These include the Crane Operator, Rigging Supervisor, Safety, Client Representative. The Client Representative, at his option, may request additional signatures including Equipment Supervisors, Safety Representatives, and the organization's Engineering Representative.

38302-11_A13.EPS

Appendix B

GROVE RT522 CRANE LOAD CHARTS

RANGE DIAGRAM

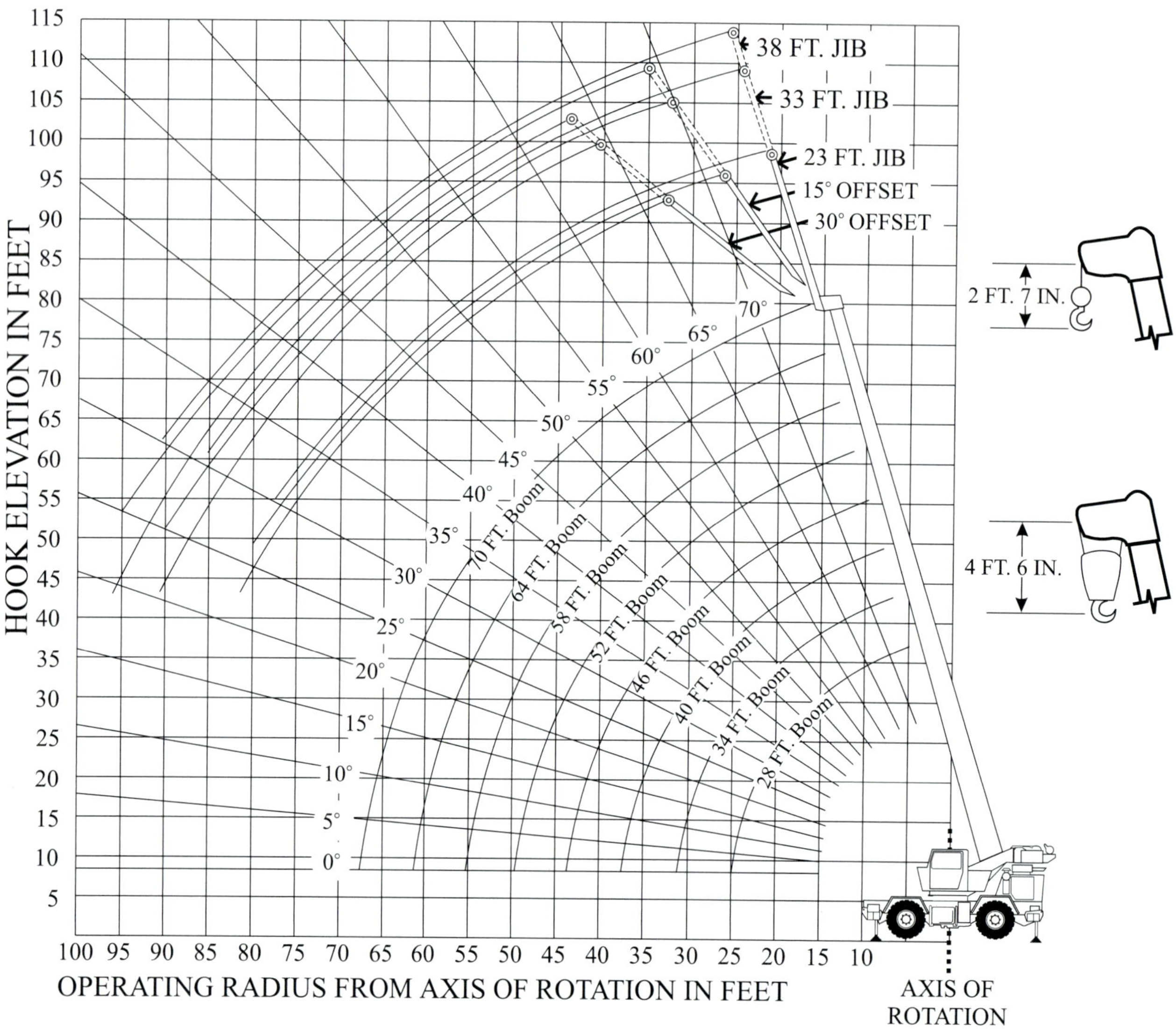

HOOK BLOCKS

22 Ton, 3 Sheave	490 lbs.
15 Ton, 2 Sheave	300 lbs.
12 Ton, 1 Sheave (15-7/8" OD)	400 lbs.
12 Ton, 1 Sheave (12-1/8" OD)	285 lbs.
Auxiliary Boom Head	100 lbs.
5-Ton Headache Ball	150 lbs.

NOTE: All load handling devices and boom attachments are considered part of the load and suitable allowances MUST BE MADE for their combined weights. Weights are for Grove furnished equipment.

WEIGHT REDUCTIONS FOR LOAD HANDLING DEVICES

23 FT. Jib WITH 28-70 FT. BOOM	
* Stowed	381 LBS.
* Erected	1,950 LBS.

** Reduction of main boom capacities.*

23-38 FT. TELE. Jib WITH 28-70 FT. BOOM	
* Stowed	604 LBS.
* Erected (Retracted)	3,659 LBS.
* Erected (Extended)	4,583 LBS.

** Reduction of main boom capacities.*

38302-11_A14.EPS

On Outriggers Fully Extended - 360°

Radius In Feet	Boom Length in Feet							
	28	34	40	46	52	58	64	70
10	44,000 (64)	36,000 (69)	36,000 (73)					
12	40,000 (59.5)	36,000 (65.5)	36,000 (70)	35,000 (73)				
15	31,000 (51.5)	31,000 (59.5)	30,700 (65)	29,850 (69)	29,150 (72)	28,600 (74.5)		
20	23,200 (36.5)	23,200 (49)	23,200 (57)	23,200 (62)	23,000 (66)	22,600 (69.5)	22,150 (72)	20,500 (74)
25	17,950 (6)	17,950 (36)	17,950 (47.5)	17,950 (54.5)	17,950 (60)	17,950 (64)	17,950 (67)	17,650 (69.5)
30		13,470 (15.5)	13,470 (36.5)	13,470 (46.5)	13,470 (53)	13,470 (58)	13,470 (62)	13,470 (65)
35			10,220 (20)	10,220 (36.5)	10,220 (45.5)	10,220 (51.5)	10,220 (56.5)	10,220 (60)
40	See Warning Note 16			8,010 (23)	8,010 (36.5)	8,010 (45)	8,010 (50.5)	8,010 (55)
45					6,530 (25)	6,530 (37)	6,530 (44.5)	6,530 (49.5)
50						5,430 (26.5)	5,430 (37)	5,430 (43.5)
55						4,440 (3.5)	4,440 (28)	4,440 (37)
60							3,620 (13)	3,620 (28.5)
65								2,980 (15.5)
Min. boom angle (deg.) for indicated length (No Load)								0
Max. boom length (Ft.) at 0 degree boom angle (No Load)								70.0

Note: Boom Angles are in degrees.

On Outriggers Fully Extended - Over Front

Radius In Feet	Boom Length in Feet							
	28	34	40	46	52	58	64	70
10	44,000 (64)	36,000 (69)	36,000 (73)					
12	40,000 (59.5)	36,000 (65.5)	36,000 (70)	35,000 (73)				
15	31,000 (51.5)	31,000 (59.5)	30,700 (65)	29,850 (69)	29,150 (72)	28,600 (74.5)		
20	23,200 (36.5)	23,200 (49)	23,200 (57)	23,200 (62)	23,000 (66)	22,600 (69.5)	22,150 (72)	20,500 (74)
25	17,950 (6)	17,950 (36)	17,950 (47.5)	17,950 (54.5)	17,950 (60)	17,950 (64)	17,950 (67)	17,650 (69.5)
30		15,350 (15.5)	15,350 (36.5)	15,350 (46.5)	15,350 (53)	15,150 (58)	14,950 (62)	14,750 (65)
35			11,900 (20)	11,900 (36.5)	11,900 (45.5)	11,900 (51.5)	11,900 (56.5)	11,900 (60)
40	See Warning Note 16			9,410 (23)	9,410 (36.5)	9,410 (45)	9,410 (50.5)	9,410 (55)
45					7,720 (25)	7,720 (37)	7,720 (44.5)	7,720 (49.5)
50						6,410 (26.5)	6,410 (37)	6,410 (43.5)
55						5,410 (3.5)	5,410 (28)	5,410 (37)
60							4,530 (13)	4,530 (28.5)
65								3,780 (15.5)
Min. boom angle (deg.) for indicated length (No Load)								0
Max. boom length (Ft.) at 0 degree boom angle (No Load)								70.0

Note: *Boom Angles are in degrees.*

38302-11_A15.EPS

14.00 x 24 Tires

Radius In Feet	Stationary Capacity	Stationary Capacity	Pick & Carry Cap. Up to 2.5 MPH
	Defined Arc (3) Over Front	360° Arc	Boom Centered (7) Over Front
10	25,600 (a)	21,100 (a)	21,610 (a)
12	21,500 (a)	15,780 (a)	18,520 (a)
15	17,770 (a)	10,430 (b)	15,050 (a)
20	12,840 (b)	5,950 (c)	12,060 (a)
25	8,480 (b)	3,570 (d)	8,480 (b)
30	6,230 (c)	2,320 (e)	6,230 (c)
35	4,650 (d)	1,470 (f)	4,650 (d)
40	3,520 (e)	910 (f)	3,000 (d)
45	2,660 (f)		2,270 (e)
50	2,020 (g)		1,680 (f)
55	1,520 (g)		1,220 (f)
60	1,130 (h)		
65	880 (h)		

16.00 x 25 Tires

Radius In Feet	Stationary Capacity	Stationary Capacity	Pick & Carry Cap. Up to 2.5 MPH
	Defined Arc (3) Over Front	360° Arc	Boom Centered (7) Over Front
10	29,150 (a)	19,640 (a)	30,460 (a)
12	24,030 (a)	14,600 (a)	26,320 (a)
15	20,150 (a)	10,390 (b)	21,670 (a)
20	13,350 (b)	5,980 (c)	13,350 (a)
25	8,610 (b)	3,500 (d)	8,290 (b)
30	6,290 (c)	2,160 (e)	6,290 (c)
35	4,650 (d)	1,430 (f)	4,650 (d)
40	3,620 (e)	810 (f)	3,620 (d)
45	2,800 (f)		2,800 (e)
50	2,120 (g)		2,120 (f)
55	1,630 (g)		1,630 (f)
60	1,240 (h)		
65	960 (h)		

20.5 x 25 Tires

Radius In Feet	Stationary Capacity	Stationary Capacity	Pick & Carry Cap. Up to 2.5 MPH
	Defined Arc Over Front (3)	360° Arc	Boom Centered Over Front (7)
10	28,380 (a)	24,450 (a)	26,020 (a)
12	23,800 (a)	17,190 (a)	22,420 (a)
15	19,100 (a)	11,640 (b)	18,380 (a)
20	13,940 (b)	7,180 (c)	13,940 (a)
25	9,160 (b)	4,330 (c)	9,160 (b)
30	6,640 (c)	2,930 (d)	4,830 (c)
35	5,100 (d)	2,000 (e)	3,790 (d)
40	4,020 (e)	1,370 (f)	3,020 (d)
45	3,140 (f)	880 (g)	2,310 (e)
50	2,430 (g)		1,710 (f)
55	1,860 (g)		1,210 (f)
60	1,440 (h)		
65	1,100 (h)		

Maximum Permissible Boom Length:	
(a) 28 Ft.	(e) 52 Ft.
(b) 34	(f) 58
(c) 40	(g) 64
(d) 46	(h) 70

		Main Boom 70 Ft.	Main Boom w/23 Ft. Jib
Front	Min. Boom Angle (deg.) for indicated Boom Length	0	0
(No Load)	Max. Boom Length (Ft.) at 0 degree Boom Angle	70	93
360°	Min. Boom Angle (deg.) for indicated Boom Length	42	51
(No Load)	Max. Boom Length (Ft.) at 0 degree Boom Angle	52	57

NOTES FOR RUBBER CAPACITIES

1. Capacities do not exceed 75% of tipping loads as determined by test in accordance with SAE J-765.
2. Capacities are applicable to machines equipped with:

	Cold Inflation	2.5 MPH
14.00 x 24 (20 ply)	115 PSI	110 PSI
16.00 x 25 (20 ply)	95 PSI	80 PSI
20.5 x 25 (20 ply)	80 PSI	65 PSI

3. Defined Arc - Over front includes " 6° on either side of longitudinal centerline of machine.
4. Capacities appearing above bold line are based on structural strength and tipping should not be relied upon as a capacity limitation.
5. Capacities are applicable only with machine on a firm level surface.
6. On-rubber lifting with jib not permitted.
7. For pick and carry operation, boom must be centered over front of machine and mechanical swing lock engaged. When handling loads in the structural range with capacities close to maximum ratings, travel should be reduced to creep speed.
8. Axle lockouts must be functioning before lifting on rubber. (Check automatic lockout system for proper functioning: Refer to "Operation and Maintenance Manual" for description of a proper functioning axle lockout system).
9. All lifting depends on proper tire inflation, capacity and condition. Capacities must be reduced for lower tire inflation pressures. See lifting capacity chart

38302-11_A16.EPS

Appendix C

MANITOWOC 4100W-2 CRANE LOAD CHARTS

MANITOWOC 4100W-2

LIFTCRANE CAPACITIES

BOOM NO. 22C WITH OPEN THROAT TOP
146,400 LB. CRANE COUNTERWEIGHT
60,000 LB. CARBODY COUNTERWEIGHT
26'6" CRAWLERS EXTENDED

WARNING: This chart will apply only when two 12,000 lb. side ctwts. and two 30,000 lb. carbody ctwts. bear MEC registered Serial Numbers.

LIFTING CAPACITIES: Capacities for various boom lengths and operating radii may be based on percent of tipping, strength of structural components, operating speeds and other factors.

Capacities are for freely suspended loads and do not exceed 75% of a static tipping load. Capacities based on structural competence are shown by shaded areas.

Capacities are shown in pounds. Deduct 1200 pounds from capacities listed when single sheave upper boom point is attached and 1500 pounds when two sheave upper boom point is attached. To comply with B30.5 requirements, upper boompoint cannot be used on the 260 ft. boom. Weight of jib, (see chart A), all load blocks, hooks, weight ball, slings, hoist lines beneath boom and jib point sheaves, etc., is considered part of the main boom load. Boom is not to be lowered beyond radii where combined weights are greater than rated capacity. Where no capacity is shown, operation is not intended or approved.

OPERATING CONDITIONS: Machine to operate in a level position on a firm surface with crawlers fully extended and gantry in working position and be rigged in accordance with and under conditions referred to in rigging drawing No. 190693 and load line specification chard No. 6592-A.

Crane operator judgement must be used to allow for dynamic load effects of swinging, hoisting or lowering, travel, as well as adverse operating conditions & physical machine depreciation.

OPERATOR RADIUS: Operating is the horizontal distance from the axis of rotation to the center of vertical hoist line or load block with the load freely suspended. Add 14" to boom point radius for radius of sheave when using single part hoist line.

Boom angle is the angle between horizontal and centerline of boom butt and inserts and is an indication of operating radius. In all cases, operating radius shall govern capacity.

BOOM POINT ELEVATION: Boom point elevation, in feet, is the vertical distance from ground level to centerline of boom point shaft.

MACHINE EQUIPMENT: Machine equipped with 26'6" extendible crawlers, 48" treads, 17' retractable gantry, 12 part boom hoist reeving, four 1 3/8" boom pendants, 1st ctwt. 41,900 lbs., 2nd ctwt. 41,500 lbs., 3rd ctwt. 39,000 lbs., two 12,000 lbs. side ctwt's. and two 30,000 lbs. carbody ctwt's.

LOAD AND WHIP LINE SPECIFICATIONS	
LOAD LINE:	1-1/8" - 6 x 31 Warrington-Seale, Extra Improved Plow Steel, Regular Lay, IWRC. Minimum Breaking Strength 65 Ton. (Approx.Weight Per Ft. in Lbs. 2.34)
WHIP LINE:	1-1/8" - Warrington-Seale, Improved Plow Steel, Regular Lay, IWRC. Minimum Breaking Strength 56.5 Ton. Maximum Load - 28,300 Lbs. Per Line. (Approx. Weight Per Ft. in Lbs. 2.34)

MAXIMUM BOOM AND JIB LENGTH LIFTED UNASSISTED				DEDUCT FROM CAPACITIES WHEN JIB IS ATTACHED	
OVER FRONT OF BLOCKED CRAWLERS		OVER SIDE OF EXTENDED CRAWLERS			
BOOM LENGTH	JIB NO. 123	BOOM LENGTH	JIB NO. 123	JIB LENGTH	JIB NO. 123
260'	--	260'	--	30'	3,000 lbs.
250'	--	250'	--	40'	3,600 lbs.
240'	40'	240'	40'	50'	4,200 lbs.
230'	60'	230'	60'	60'	4,900 lbs.
Load block, hook and weight ball on ground to start.					

FOR JIB CAPACITIES, CONSULT JIB CHART.

HOIST REEVING FOR MAIN LOAD BLOCK						
No.Parts Of Line	1	2	3	4	5	6
Max Load - Lbs.	32,500	65,000	97,500	130,000	162,500	195,000
No. Parts of Line	7	8	9	10	11	12
Max. Load - Lbs.	227,500	260,000	292,500	325,000	357,500	400,000
No. Parts of Line	13					
Max. Load - Lbs.	430,000					

BOOM LGTH FEET	OPER. RAD. FEET	BOOM ANG. DEG.	BOOM POINT ELEV.	CAPACITY: CRAWLERS EXTENDED
70	16.5	79.7	75.9	460,000
	17	79.3	75.8	400,000
	18	78.5	75.6	380,100
	19	77.6	75.4	363,000
	20	76.8	75.1	347,300
	22	75.1	74.6	319,600
	24	73.4	74.1	293,400
	26	71.7	73.5	266,100
	28	69.9	72.8	237,500
	30	68.2	72.0	214,300
	32	66.4	71.2	195,100
	34	64.6	70.2	178,900
	36	62.8	69.3	165,200
	38	60.9	68.2	153,300
	40	59.1	67.0	143,000
	45	54.1	63.7	122,100
	50	48.9	59.8	106,300
	55	43.2	54.9	93,900
	60	36.9	49.0	84,000
	65	29.4	41.3	75,800
	70	19.5	30.3	63,900

BOOM LGTH FEET	OPER. RAD. FEET	BOOM ANG. DEG.	BOOM POINT ELEV.	CAPACITY: CRAWLERS EXTENDED
80	17	80.6	85.9	392,800
	18	79.9	85.8	378,900
	19	79.2	85.6	361,800
	20	78.5	85.4	346,100
	22	77.0	84.9	318,400
	24	75.5	84.5	292,500
	26	74.0	83.9	265,600
	28	72.5	83.3	237,000
	30	71.0	82.7	213,800
	32	69.5	81.9	194,600
	34	68.0	81.2	178,500
	36	66.4	80.3	164,700
	38	64.8	79.4	152,800
	40	63.3	78.4	142,400
	45	59.2	75.7	121,500
	50	54.9	72.5	105,700
	55	50.4	68.6	93,300
	60	45.6	64.1	83,400
	65	40.3	58.8	75,200
	70	34.4	52.2	68,300
	75	27.4	43.9	52,500
	80	18.2	32.0	53,900

BOOM LGTH FEET	OPER. RAD. FEET	BOOM ANG. DEG.	BOOM POINT ELEV.	CAPACITY: CRAWLERS EXTENDED
90	18	81.1	95.9	355,400
	19	80.4	95.7	346,900
	20	79.8	95.6	336,900
	22	78.5	95.2	317,400
	24	77.2	94.7	291,700
	26	75.9	94.3	264,800
	28	74.5	93.7	236,600
	30	73.2	93.2	213,400
	32	71.9	92.5	194,200
	34	70.5	91.9	178,000
	36	69.2	91.1	164,200
	38	67.8	90.3	152,300
	40	66.4	89.5	142,000
	45	62.9	87.1	121,100
	50	59.3	84.4	105,200
	55	55.5	81.2	92,800
	60	51.5	77.5	82,900
	65	47.3	73.2	74,700
	70	42.8	68.2	67,800
	75	37.9	62.3	62,000
	80	32.4	55.2	57,000
	85	25.8	46.2	52,600
	90	17.1	33.5	45,900

BOOM LGTH FEET	OPER. RAD. FEET	BOOM ANG. DEG.	BOOM POINT ELEV.	CAPACITY: CRAWLERS EXTENDED
100	19	81.4	105.9	332,900
	20	80.8	105.7	327,100
	22	79.6	105.4	316,200
	24	78.5	105.0	290,800
	26	77.3	104.5	263,900
	28	76.1	104.1	236,200
	30	74.9	103.6	212,900
	32	73.7	103.0	193,700
	34	72.5	102.4	177,500
	36	71.3	101.7	163,700
	38	70.1	101.0	151,800
	40	68.9	100.3	141,400
	45	65.8	98.2	120,500
	50	62.6	95.8	104,700
	55	59.3	93.0	92,300
	60	55.9	89.8	82,300
	65	52.4	86.2	74,100
	70	48.7	82.1	67,200
	75	44.8	77.4	61,400
	80	40.5	72.0	56,400
	85	35.9	65.6	52,000
	90	30.7	58.0	48,200
	95	24.5	48.5	44,900
	100	16.3	35.0	39,300

CAUTION! CHECK AMOUNT OF COUNTERWEIGHT ON MACHINE BEFORE USE OF THIS CHART

38302-11_A17.EPS

LIFTCRANE CAPACITIES – 4100W SERIES 2 BOOM NO 22C WITH OPEN-THROAT TOP, CONTINUED

BOOM LGTH FEET	OPER. RAD. FEET	BOOM ANG. DEG.	BOOM POINT ELEV.	CAPACITY: CRAWLERS EXTENDED
110	22	82.3	115.7	240,500
	24	81.3	115.4	232,100
	26	80.2	115.0	224,400
	28	79.1	114.6	217,300
	30	78.1	114.1	210,700
	32	77.0	113.6	193,300
	34	75.9	113.1	177,100
	36	74.8	112.5	163,300
	38	73.7	111.8	151,400
	40	72.6	111.2	141,000
	45	69.9	109.3	120,100
	50	67.0	107.2	104,200
	55	64.1	104.7	91,800
	60	61.2	101.9	81,800
	65	58.1	98.8	73,600
	70	54.9	95.3	66,800
	75	51.6	91.3	60,900
	80	48.1	86.8	55,900
	85	44.4	81.7	51,600
	90	40.4	75.9	47,800
	95	36.1	69.2	44,400
	100	31.1	61.1	41,400
	105	25.3	51.1	38,700
	110	17.6	37.2	33,900
120	22	83.0	125.9	228,700
	24	82.0	125.5	220,500
	26	81.0	125.2	213,000
	28	80.0	124.8	206,100
	30	79.1	124.4	199,700
	32	78.1	123.9	192,800
	34	77.1	123.4	176,600
	36	76.1	122.9	162,800
	38	75.1	122.3	150,900
	40	74.1	121.7	140,500
	45	71.6	120.0	119,500
	50	69.1	118.1	103,700
	55	66.5	115.9	91,300
	60	63.8	113.4	81,300
	65	61.1	110.6	73,100
	70	58.3	107.5	66,200
	75	55.4	104.0	60,400
	80	52.4	100.1	55,300
	85	49.2	95.8	51,000
	90	45.9	91.0	47,200
	95	42.4	85.5	43,800
	100	38.6	79.4	40,800
	105	34.4	72.2	38,100
	110	29.7	63.7	35,700
	115	24.1	53.2	33,500
130	24	82.6	135.7	212,700
	26	81.7	135.3	205,300
	28	80.8	135.0	198,600
	30	79.9	134.6	192,300
	32	79.0	134.2	186,500
	34	78.1	133.7	176,300
	36	77.2	133.2	162,500
	38	76.3	132.7	150,600
	40	75.4	132.1	140,200
	45	73.1	130.6	119,300
	50	70.8	128.8	103,400
	55	68.4	126.8	91,000
	60	66.0	124.6	81,000
	65	63.5	122.0	72,800
	70	61.0	119.2	65,900
	75	58.4	116.1	60,000
	80	55.7	112.7	55,000
	85	53.0	108.9	50,700
	90	50.1	104.8	46,800
	95	47.1	100.1	43,500
	100	43.9	95.0	40,500
	105	40.6	89.2	37,800
	110	36.9	82.6	35,300
	115	33.0	75.1	33,100
	120	28.5	66.1	31,100
	125	23.1	55.1	29,300
140	26	82.3	145.5	198,400
	28	81.5	145.1	191,800
	30	80.7	144.8	185,600
	32	79.8	144.4	179,900
	34	79.0	144.0	174,700
	36	78.1	143.5	162,000
	38	77.3	143.0	150,100
	40	76.5	142.5	139,700
	45	74.3	141.1	118,700
	50	72.2	139.4	102,800
	55	70.0	137.6	90,400
	60	67.8	135.5	80,400
	65	65.6	133.2	72,200
	70	63.3	130.7	65,300
	75	60.9	127.9	59,500
	80	58.5	124.8	54,400
	85	56.1	121.4	50,100
	90	53.5	117.7	46,200
	95	50.9	113.7	42,900
	100	48.1	109.2	39,900
140	105	45.3	104.2	37,200
	110	42.2	98.8	34,700
	115	39.0	92.7	32,500
	120	35.5	85.7	30,500
	125	31.7	77.8	28,700
	130	27.4	68.5	27,000
	135	22.2	57.0	25,500
150	28	82.1	155.3	185,600
	30	81.3	154.9	179,600
	32	80.5	154.6	174,000
	34	79.7	154.2	168,800
	36	79.0	153.8	161,600
	38	78.2	153.3	149,600
	40	77.4	152.8	139,200
	45	75.4	151.5	118,200
	50	73.4	150.0	102,400
	55	71.4	148.3	89,900
	60	69.4	146.4	79,900
	65	67.3	144.3	71,700
	70	65.2	141.9	64,800
	75	63.0	139.4	59,000
	80	60.9	136.6	53,900
	85	58.6	133.5	49,600
	90	56.3	130.1	45,700
	95	54.0	126.5	42,400
	100	51.5	122.5	39,400
	105	49.0	118.2	36,700
	110	46.4	113.4	34,200
	115	43.6	108.2	32,000
	120	40.7	102.4	30,000
	125	37.6	96.0	28,200
	130	34.2	88.8	26,500
	135	30.5	80.5	25,000
	140	26.4	70.8	23,500
	145	21.4	58.8	22,200
160	28	82.6	165.4	179,800
	30	81.8	165.1	173,900
	32	81.1	164.7	168,400
	34	80.4	164.4	163,300
	36	79.7	164.0	158,600
	38	78.9	163.6	149,100
	40	78.2	163.1	138,700
	45	76.3	161.9	117,700
	50	74.5	160.5	101,800
	55	72.6	158.9	89,300
	60	70.7	157.1	79,300
	65	68.8	155.1	71,100
	70	66.8	153.0	64,200
	75	64.9	150.6	58,400
	80	62.9	148.0	53,300
	85	60.8	145.2	49,000
	90	58.7	142.2	45,100
	95	56.6	138.8	41,800
	100	54.4	135.2	38,700
	105	52.1	131.4	36,000
	110	49.7	127.1	33,600
	115	47.3	122.5	31,400
	120	44.8	117.5	29,400
	125	42.1	112.0	27,600
	130	39.3	105.9	25,900
	135	36.3	99.2	24,300
	140	33.1	91.7	22,900
	145	29.5	83.0	21,600
	150	25.5	72.9	20,400
	155	20.7	60.5	19,200
170	30	82.3	175.2	169,000
	32	81.6	174.9	163,600
	34	81.0	174.5	158,700
	36	80.3	174.2	154,000
	38	79.6	173.8	148,800
	40	78.9	173.3	138,400
	45	77.2	172.2	117,400
	50	75.4	170.9	101,500
	55	73.7	169.4	89,100
	60	71.9	167.7	79,000
	65	70.1	165.9	70,800
	70	68.3	163.9	63,900
	75	66.4	161.7	58,100
	80	64.6	159.3	53,000
	85	62.7	156.7	48,700
	90	60.8	153.9	44,900
	95	58.8	150.8	41,500
	100	56.8	147.5	38,500
	105	54.7	144.0	35,800
	110	52.6	140.2	33,300
	115	50.4	136.0	31,100
	120	48.1	131.5	29,100
	125	45.8	126.7	27,300
	130	43.3	121.4	25,600
	135	40.8	115.7	24,100
	140	38.1	109.3	22,600
	145	35.2	102.3	21,300
	150	32.0	94.5	20,100
	155	28.6	85.5	18,900
	160	24.7	75.1	17,800
	165	20.0	62.2	16,400
180	32	82.1	185.0	158,700
	34	81.5	184.7	154,000
	36	80.8	184.3	149,500
	38	80.2	184.0	145,200
	40	79.5	183.6	137,900
	45	77.9	182.5	116,900
	50	76.3	181.2	101,000
	55	74.6	179.8	88,500
	60	72.9	178.3	78,500
	65	71.3	176.6	70,300
	70	69.6	174.7	63,400
	75	67.8	172.6	57,500
	80	66.1	170.4	52,500
	85	64.3	168.0	48,100
	90	62.5	165.4	44,300
	95	60.7	162.5	40,900
	100	58.9	159.5	37,900
	105	57.0	156.2	35,200
	110	55.0	152.7	32,700
	115	53.0	149.0	30,500
	120	51.0	144.9	28,500
	125	48.9	140.5	26,700
	130	46.7	135.8	25,000
	135	44.4	130.7	23,500
	140	42.0	125.2	22,000
	145	39.6	119.2	20,700
	150	36.9	112.6	19,500
	155	34.1	105.3	18,300
	160	31.1	97.2	17,200
	165	27.7	87.9	16,200
	170	23.9	77.1	15,300
	175	19.4	63.8	13,500
190	32	82.5	195.1	153,100
	34	81.9	194.8	148,800
	36	81.3	194.5	144,800
	38	80.7	194.1	140,800
	40	80.1	193.8	137,000
	45	78.5	192.7	116,400
	50	77.0	191.5	100,500
	55	75.4	190.2	88,000
	60	73.9	188.8	78,000
	65	72.3	187.1	69,700
	70	70.7	185.4	62,800
	75	69.1	183.4	57,000
	80	67.4	181.4	52,000
	85	65.8	179.1	47,600
	90	64.1	176.6	43,800
	95	62.4	174.0	40,400
	100	60.7	171.2	37,400
	105	58.9	168.2	34,700
	110	57.1	164.9	32,200
	115	55.3	161.4	30,000
	120	53.4	157.7	28,000
	125	51.5	153.7	26,200
	130	49.5	149.5	24,500
	135	47.5	144.9	22,900
	140	45.3	140.0	21,500
	145	43.1	134.6	20,200
	150	40.8	128.9	18,900
	155	38.4	122.6	17,800
	160	35.9	115.8	16,700
	165	33.2	108.3	15,700
	170	30.2	99.8	14,800
	175	27.0	90.3	13,900
	180	23.3	79.1	13,100
	185	18.9	65.4	11,000
200	34	82.3	204.9	143,100
	36	81.7	204.6	139,500
	38	81.2	204.3	136,000
	40	80.6	203.9	132,500
	45	79.1	203.0	115,900
	50	77.7	201.8	100,000
	55	76.2	200.6	87,500
	60	74.7	199.2	77,400
	65	73.2	197.7	69,200
	70	71.7	196.0	62,300
	75	70.2	194.2	56,400
	80	68.6	192.2	51,400
	85	67.1	190.1	47,000
	90	65.5	187.8	43,200
	95	63.9	185.3	39,800
	100	62.3	182.7	36,800
	105	60.6	179.8	34,100
	110	59.0	176.8	31,600
	115	57.3	173.6	29,400
	120	55.5	170.2	27,400
	125	53.0	166.5	25,600
	130	51.9	162.6	23,900
	135	50.1	158.4	22,300
	140	48.1	153.9	20,900
	145	46.2	149.1	19,600
	150	44.1	144.0	18,300
	155	42.0	138.4	17,200
	160	39.7	132.5	16,100
	165	37.4	126.0	15,100
	170	34.9	118.9	14,200
200	175	32.3	111.1	13,300
	180	29.4	102.4	12,400
	185	26.2	92.5	11,700
	190	22.6	81.0	10,900
	195	18.4	67.0	8,500
210	36	82.1	214.7	134,200
	38	81.6	214.4	131,000
	40	81.0	214.1	127,900
	45	79.6	213.2	115,500
	50	78.3	212.1	99,600
	55	76.9	210.9	87,100
	60	75.4	209.6	77,100
	65	74.0	208.1	68,800
	70	72.6	206.6	61,900
	75	71.2	204.8	56,100
	80	69.7	203.0	51,000
	85	68.2	201.0	46,700
	90	66.7	198.8	42,800
	95	65.2	196.5	39,400
	100	63.7	194.0	36,400
	105	62.2	191.3	33,700
	110	60.6	188.5	31,300
	115	59.0	185.5	29,100
	120	57.4	182.3	27,100
	125	55.7	178.9	25,200
	130	54.1	175.2	23,500
	135	52.3	171.4	22,000
	140	50.6	167.3	20,600
	145	48.8	162.9	19,200
	150	46.9	158.2	18,000
	155	45.0	153.2	16,800
	160	43.0	147.9	15,800
	165	40.9	142.1	14,800
	170	38.7	135.9	13,800
	175	36.5	129.2	12,900
	180	34.0	121.9	12,100
	185	31.5	113.9	11,300
	190	28.7	104.9	10,600
	195	25.6	94.7	9,900
	200	22.1	82.9	8,800
	205	17.9	68.5	6,500
220	38	82.0	224.6	125,700
	40	81.4	224.2	122,800
	45	80.1	223.3	115,000
	50	78.8	222.3	99,100
	55	77.5	221.2	86,600
	60	76.1	220.0	76,600
	65	74.8	218.6	68,300
	70	73.4	217.1	61,400
	75	72.0	215.4	55,500
	80	70.7	213.7	50,500
	85	69.3	211.8	46,100
	90	67.9	209.7	42,300
	95	66.4	207.5	38,900
	100	65.0	205.2	35,800
	105	63.6	202.7	33,100
	110	62.1	200.0	30,700
	115	60.6	197.2	28,500
	120	59.1	194.2	26,500
	125	57.5	191.0	24,600
	130	55.9	187.6	23,000
	135	54.3	184.0	21,400
	140	52.7	180.2	20,000
	145	51.0	176.1	18,600
	150	49.3	171.8	17,400
	155	47.6	167.3	16,300
	160	45.7	162.4	15,200
	165	43.9	157.2	14,200
	170	41.9	151.7	13,200
	175	39.9	145.7	12,300
	180	37.8	139.3	11,500
	185	35.6	132.4	10,700
	190	33.2	124.8	10,000
	195	30.7	116.6	9,300
	200	28.0	107.3	8,600
	205	25.0	96.9	8,000
	210	21.5	84.8	6,500

CAPACITIES CONTINUED ON NEXT PAGE

These load charts are intended for instructional purposes only. They were derived from manufacturer sales information which may not be complete or machine specific. Not responsible for typographical errors.

38302-11_A18.EPS

LIFTCRANE CAPACITIES – 4100W SERIES 2 BOOM NO 22C WITH OPEN-THROAT TOP, CONTINUED

SEE CONDITIONS ON FRONT PAGE

BOOM LGTH FEET	OPER. RAD. FEET	BOOM ANG. DEG.	BOOM POINT ELEV.	CAPACITY: CRAWLERS EXTENDED
230	38	81.5	234.5	132,900
	40	81.0	234.2	125,200
	45	79.7	233.3	114,500
	50	78.5	232.4	98,600
	55	77.2	231.3	86,100
	60	75.9	230.1	76,100
	65	74.6	228.8	67,800
	70	73.3	227.3	60,900
	75	72.0	225.8	55,000
	80	70.7	224.1	50,000
	85	69.4	222.3	45,600
	90	68.0	220.3	41,700
	95	66.7	218.2	38,400
	100	65.3	216.0	35,300
	105	64.0	213.6	32,600
	110	62.6	211.1	30,200
	115	61.1	208.4	28,000
	120	59.7	205.6	26,000
	125	58.3	202.6	24,100
	130	56.8	199.4	22,400
	135	55.3	196.0	20,900
	140	53.8	192.5	19,400
	145	52.2	188.7	18,100
	150	50.6	184.7	16,900
	155	49.0	180.5	15,700
	160	47.3	176.0	14,700
	165	45.6	171.3	13,600
	170	43.8	166.2	12,700
	175	42.0	160.8	11,800
	180	40.1	155.1	11,000
	185	38.1	148.9	10,200
	190	36.0	142.3	9,400
	195	33.9	135.1	8,700
	200	31.6	127.3	8,100
	205	29.1	118.8	7,400
	210	26.4	109.3	6,800
240	40	81.4	244.3	123,400
	45	80.2	243.5	112,600
	50	78.9	242.6	98,100
	55	77.7	241.5	85,600
	60	76.5	240.4	75,600
	65	75.3	239.1	67,300
	70	74.0	237.7	60,400
	75	72.8	236.3	54,500
	80	71.5	234.6	49,500
	85	70.3	232.9	45,100
	90	69.0	231.1	41,200
	95	67.7	229.1	37,800
	100	66.4	227.0	34,800
	105	65.1	224.7	32,100
	110	63.8	222.3	29,700
	115	62.5	219.8	27,400
	120	61.1	217.1	25,400
	125	59.7	214.3	23,600
	130	58.3	211.3	21,900
	135	56.9	208.1	20,300
	140	55.5	204.7	18,900
	145	54.0	201.2	17,600
	150	52.5	197.5	16,300
	155	51.0	193.5	15,200
	160	49.5	189.4	14,100
	165	47.9	185.0	13,100
	170	46.2	180.3	12,200
	175	44.6	175.4	11,300
	180	42.8	170.2	10,400
	185	41.0	164.6	9,600
	190	39.2	158.7	8,900
	195	37.3	152.3	8,200
	200	35.2	145.5	7,500
	205	33.1	138.1	6,700
	210	30.9	130.1	6,000
250	45	80.6	253.6	106,600
	50	79.4	252.7	97,800
	55	78.2	251.7	85,200
	60	77.1	250.6	75,200
	65	75.9	249.4	66,900
	70	74.7	248.1	60,000
	75	73.5	246.7	54,100
	80	72.3	245.2	49,100
	85	71.1	243.5	44,700
	90	69.9	241.7	40,800
	95	68.7	239.8	37,400
	100	67.4	237.8	34,400
	105	66.2	235.7	31,700
	110	64.9	233.4	29,300
	115	63.6	231.0	27,100
	120	62.4	228.5	25,000
	125	61.1	225.8	23,200
	130	59.7	222.9	21,500
	135	58.4	219.9	20,000
	140	57.0	216.8	18,500
	145	55.7	213.4	17,200
	150	54.3	209.9	16,000
	155	52.8	206.2	14,800
	160	51.4	202.4	13,700
	165	49.9	198.3	12,700
	170	48.4	193.9	11,800
	175	46.8	189.4	10,900
	180	45.3	184.5	10,100
	185	43.6	179.4	9,300
	190	41.9	174.0	8,500
	195	40.2	168.3	7,800
	200	38.4	162.2	7,000
	205	36.5	155.7	6,200
	210	34.5	148.6	5,500
260	45	80.9	263.7	104,800
	50	79.8	262.9	95,700
	55	78.7	261.9	84,700
	60	77.6	260.9	74,700
	65	76.4	259.7	66,400
	70	75.3	258.5	59,500
	75	74.2	257.1	53,600
	80	73.0	255.6	48,500
	85	71.8	254.1	44,100
	90	70.7	252.4	40,300
	95	69.5	250.6	36,900
	100	68.3	248.6	33,900
	105	67.1	246.6	31,100
	110	65.9	244.4	28,700
	115	64.7	242.1	26,500
	120	63.5	239.7	24,500
	125	62.3	237.1	22,600
	130	61.0	234.4	20,900
	135	59.7	231.6	19,400
	140	58.5	228.6	17,900
	145	57.2	225.4	16,600
	150	55.8	222.1	15,400
	155	54.5	218.7	14,200
	160	53.1	215.0	13,100
	165	51.7	211.2	12,100
	170	50.3	207.1	11,200
	175	48.9	202.9	10,300
	180	47.4	198.4	9,500
	185	45.9	193.7	8,700
	190	44.3	188.7	7,900
	195	42.7	183.4	7,000
	200	41.1	177.8	6,200
	205	39.4	171.9	5,400
	210	37.6	165.6	4,700

38302-11_A19.EPS

DESCRIPTION	APPROX. WEIGHT (IN LBS.)
JIB NO. 123	
Jib Top - 15' (w/Jib point)	695
Jib Butt - 15'	690
Jib Insert - 10'	340
Basic Pendant - 33' 3-3/4" (2 Req'd)	115 each
Pendant - 10' (2 Per insert)	65 each
Jib Backstay Pendant	155 each
Jib Strut - 12'- 6"	365
COMPONENTS	
Hook Rollers (6) - w/shafts	1,020
Light Plant - 6.5KW - w/mounting platform	1,390
Catwalk - left and right side w/rails	1,320
Lagging - 27-5/8" dia. plain	1,410
Boom Hoist Rope - 12 Part - 760' of 7/8" - 6 x 26	1,080
Wire Rope Guide Assembly - Lower	325
Wire Rope Guide Assembly - Upper	510
Rope Guide Roller Assembly	55 each
2-Part Gantry w/Telescopic Backhitch	7,805
Equalizer	2,000
Hoist Line - 1-1/8" - 6 x 31	2.34 lbs./ft.
Whip Line - 1-1/8" - 6 x 31	2.34 lbs./ft.
15-Ton Hook and Weight Ball	865
100-Ton Hook Block Assembly	2,065
200-Ton Hook Block Assembly	4,900
230-Ton Hook Block Assembly	5,375
Boom Stop - Telescopic Air Cushioned	675
Dragline Fairlead - Revolving	1,910
Dragline Fairlead - Hinged	9,330

NOTE: The above weights may fluctuate up or down 5% due to manufacturing tolerances.

38302-11_A20.EPS

Additional Resources

This module presents thorough resources for task training. The following resource material is suggested further study.

Bob's Rigging and Crane Handbook, Latest Edition. Leawood, KS: Pellow Engineering Services.

IPT's Crane and Rigging Handbook, Latest Edition. Edmonton, Alberta, Canada: IPT Publishing and Training.

Figure Credits

Manitowoc Crane Group, Figures 1-3, 39, and Appendix C

Link-Belt Construction Equipment Co., Figures 14-16, 27, 29, 32, and 35

Terex, Figures 17 and 18

Grove, Figures 19, 20, 22–26, 33, 36–38, and Appendix B

Essex Crane Rental Corp., **www.essexcrane.com**, Figure 28

The Crosby Group, Inc., Table 1

Crane Inspection & Certification Bureau, Appendix A

NCCER CURRICULA — USER UPDATE

NCCER makes every effort to keep its textbooks up-to-date and free of technical errors. We appreciate your help in this process. If you find an error, a typographical mistake, or an inaccuracy in NCCER's curricula, please fill out this form (or a photocopy), or complete the online form at **www.nccer.org/olf**. Be sure to include the exact module ID number, page number, a detailed description, and your recommended correction. Your input will be brought to the attention of the Authoring Team. Thank you for your assistance.

Instructors – If you have an idea for improving this textbook, or have found that additional materials were necessary to teach this module effectively, please let us know so that we may present your suggestions to the Authoring Team.

NCCER Product Development and Revision

13614 Progress Blvd., Alachua, FL 32615

Email: curriculum@nccer.org
Online: www.nccer.org/olf

❑ Trainee Guide ❑ AIG ❑ Exam ❑ PowerPoints Other ______________________________

Craft / Level: Copyright Date:

Module ID Number / Title:

Section Number(s):

Description:

Recommended Correction:

Your Name:

Address:

Email: Phone:

Glossary

Aeolian vibration: A vibration resulting from smooth airflow across a cylindrical surface, such as a conductor. The airflow creates a vortex that alternates between the top and bottom surfaces of the conductor, creating a vertical vibration that can cause metal fatigue and abrasion damage to the conductor. Turbulent airflow results in less aeolian vibration due to the inconsistent frequency of the vortices.

Boom point elevation: Distance from the ground to the center of the boom point sheaves.

Carbon tracking: The deposition of carbon particles on a surface that indicates a path of current flow is present or has been active.

Engineering: Practical application of the knowledge of physics, leverage, and gravity as they relate to lifting operations.

Equalizer: Any of the various devices or appliances for equalizing stresses, strains, and pressures.

Ergonomic component: A distance value that is added to the electrical component of the minimum approach distance to reach the MAD values shown in *OSHA Standard 1910.269(l)(10), Tables R-6 through R-9*. For voltages up to 72.5kV, the ergonomic component is 2 feet; for voltages above 72.5kV, the ergonomic component is 1 foot.

Hold off: Term describing the request by linemen for any automatic reclosers related to the section of line they are working on to be disabled or set to a non-automatic position; also known as a one-shot request.

Interpolation: Method of estimating or finding an intermediate value or term in a sequence or between two known values.

Lift plan: A written plan that outlines all weight, rigging, and configuration considerations for a lift.

Sheave: A grooved pulley used to guide a rope.

Strain carrier: An assembly of FRP strain poles and yokes designed to take the place of an insulator or similar component in holding or restraining a conductor.

Strain link stick: An FRP pole with a gated hook or roller on one end that is used to provide electrical isolation from energized components and conductors when connecting strap hoists, pulling grips, and similar tools.

Trunnions: Special nuts with long threads that connect strain poles to the back (cold) yoke while allowing some freedom for the poles to pivot. Adjusting the trunnions lengthens or shortens the strain carrier.

Index